D1153233

GLOBALIZATION OF TECHNOLOGY

INTERNATIONAL PERSPECTIVES

Proceedings of the Sixth Convocation of
The Council of Academies of
Engineering and Technological Sciences

Janet H. Muroyama and H. Guyford Stever, *Editors*

NATIONAL ACADEMY PRESS
Washington, D.C. 1988

National Academy Press 2101 Constitution Avenue, NW Washington, DC 20418

As host of the Sixth Convocation of the Council of Academies of Engineering and Technological Sciences, the National Academy of Engineering, through the National Academy Press, was responsible for publication of the proceedings volume. The National Academy of Engineering was established in 1964, under the charter of the National Academy of Sciences, as a parallel organization of outstanding engineers. It is autonomous in its administration and in the selection of its members, sharing with the National Academy of Sciences the responsibility for advising the federal government. The National Academy of Engineering also sponsors engineering programs aimed at meeting national needs, encourages education and research, and recognizes the superior achievement of engineers. Dr. Robert M. White is president of the National Academy of Engineering.

This publication has been reviewed by a group other than the authors according to procedures approved by a Report Review Committee. The interpretations and conclusions in this publication are those of the authors and do not purport to represent the views of the council, officers, or staff of either the National Academy of Engineering or the Council of Academies of Engineering and Technological Sciences.

The Sixth Convocation was supported by the Council of Academies of Engineering and Technological Sciences. Funds for the convocation and the publication of the proceedings volume were provided by the National Academy of Engineering's Technology Agenda Program.

Library of Congress Cataloging-in-Publication Data

Council of Academies of Engineering and Technological Sciences.
Convocation (6th : 1987: Washington, D.C.)
Globalization of technology : international perspectives :
proceedings of the Sixth Convocation of the Council of Academies of
Engineering and Technological Sciences / Janet H. Muroyama
and H. Guyford Stever, editors.
p. cm.
Includes index.
ISBN 0-309-03843-x. ISBN 0-309-03842-1 (pbk.)
1. Technology—Congresses. I. Muroyama, Janet H. II. Stever, H.
Guyford, III. Title.
T6.C55 1987
600—dc19 88-12090
CIP

Printed in the United States of America

Preface and Acknowledgments

TECHNOLOGY'S HIGHEST PURPOSE lies in helping to bring about a better world, that is, in creating systems that are a benefit to society. Historically, technology has enabled nations to expand the economic pie so that the standard of living for all is increased. The globalization of the world economy makes it imperative that this aspiration be achieved through increased international cooperation, marking a permanent transformation in the engineering endeavor.

This change is already evident in many areas. No longer does a small group of nations have a monopoly on technological superiority, for centers of excellence in electronics, materials, propulsion, manufacturing, and biotechnology exist in many countries. Increased cross-national investment has enabled engineers from one country to work for firms from other countries. Vast communication networks among interdependent firms and research centers facilitate the flow of information about technological developments, creating an impact far beyond the borders of the country in which the discovery originated. New cooperative arrangements such as joint ventures, purchases of licenses and patents, research consortia, and branch laboratories have been established to promote technical exchange. Standards, certification, and product design are being harmonized as the products of engineering become increasingly global in reach and application.

The Sixth Convocation of the Council of Academies of Engineering and Technological Sciences, held March 30 through April 1, 1987, presented a unique opportunity for engineers and technologists from around the world to gather and exchange ideas on the interaction between technological advance and the global economy. In addition to facilitating provocative discussion, the event was noteworthy because the Council is a singular nongovernmental, multilateral mechanism for bringing the leaders of the world's engineering community together. This book contains the papers presented at that convocation.

It is my hope that the Council of Academies of Engineering and Technological Sciences will gain wider recognition for its activities on behalf of international engineering exchange. By providing forums for discussion of technological issues and by acting as a stimulus for the formation of national engineering academies, the Council plays an important role in a world where technological cooperation has become a necessity.

On behalf of the National Academy of Engineering, which hosted the Sixth Convocation, I would like to thank the officers of the council member academies who helped to design the convocation; H. Guyford Stever, who chaired the meeting; and Simon Ramo, our keynote speaker. I would like to extend special thanks to the session chairmen—Alf Åkerman, Marco A. Murray-Lasso, Morris Tanenbaum, Sir Francis Tombs, and Sir David Zeidler—and the speakers (listed on page 191) who contributed so much to the substance of the proceedings. Steven Anastasion, Janet Muroyama, and Benta Sims deserve special mention for their efforts in organizing the convocation. For their work in preparing the manuscript for publication, I would like to thank Janet Muroyama, Jesse Ausubel, Dale Langford, Annette McNeil, Caroline Anderson, and Barbara Becker.

ROBERT M. WHITE
President
National Academy of Engineering

Contents

GLOBALIZATION OF TECHNOLOGY

Overview

H. GUYFORD STEVER AND JANET H. MUROYAMA

THE EFFECTS OF TECHNOLOGICAL CHANGE on the global economic structure are creating immense transformations in the way companies and nations organize production, trade goods, invest capital, and develop new products and processes. Sophisticated information technologies permit instantaneous communication among the far-flung operations of global enterprises. New materials are revolutionizing sectors as diverse as construction and communications. Advanced manufacturing technologies have altered long-standing patterns of productivity and employment. Improved air and sea transportation has greatly accelerated the worldwide flow of people and goods.

All this has both created and mandated greater interdependence among firms and nations. The rapid rate of innovation and the dynamics of technology flows mean that comparative advantage is short-lived. To maximize returns, arrangements such as transnational mergers and shared production agreements are sought to bring together partners with complementary interests and strengths. This permits both developed and developing countries to harness technology more efficiently, with the expectation of creating higher standards of living for all involved.

Rapid technological innovation and the proliferation of transnational organizations are driving the formation of a global economy that sometimes conflicts with nationalistic concerns about maintaining comparative advantage and competitiveness. It is indeed a time of transition for firms and governments alike. This book provides a broad overview of these issues and seeks to shed light on such areas as the changing nature of international competition, influences of new technologies on international trade, and economic and social concerns arising from differences in national cultures and standards of living associated with adoption and use of new technologies.

The volume is a compilation of papers presented at the Sixth Convocation of the Council of Academies of Engineering and Technological Sciences held in Washington, D.C. in the spring of 1987. The convocation brought together about one hundred leaders in technology from more than twenty countries to discuss issues of "Technology and the Global Economy." The program of the convocation was structured around four objectives:

- identification and discussion of the driving technologies of the current era, for example, in materials, information, and manufacturing;
- evaluation of how technological advances are transforming industrial sectors such as telecommunications and construction;
- exploration of how in turn the global economy is affecting technology and production through such factors as marketing strategies, intellectual property rights, and financial markets; and
- clarification of regional and national consequences of globalizing industries for several geographical areas including the Pacific Rim, Western Europe, and Latin America.

An overall assessment of the issues raised was provided in conclusion by a panel consisting of Morris Tanenbaum, Wolf Häfele, Sir Robin Nicholson, and Robert Malpas. On the one hand, their assessment made clear that though most technological advance occurs in industry, there are too few mechanisms for exchange of views on international technology and cooperation that involve both private and public sector representatives in a forum not constrained by the formal policies and stands of national governments. There is great need for improved and more open lines of international communication on topics where engineering and technology intertwine with trade and economic growth.

At the same time, the panelists' evaluations made clear a hierarchy of four sets of relationships among technology, technologists, and the societies they attempt to serve. The first of these includes relationships at the human level, ranging from professional education to relations between management and labor to the public's understanding of the impact of technology on our lives. The second includes relationships at the institutional level, that is, the impact of technology on the management of businesses and industries. The third relationship is at the national level, where public and private interactions determine the use of technology and possibly a country's ability to grow economically. The fourth relationship occurs at the international level. Here information flows, trade frictions, and alliances characterize technological development, its diffusion, global competition, and economic advance.

At the human level a key area of change is the invisible contract between a manufacturing company and its customers and employees. In the factory, we are seeing a movement away from the expectation that workers should be organized to fit the technologies and a movement toward networking and small teams. Wolf Häfele referred to this as the evolution of a higher level of integration between technology and human relations. This is evident, he said, in the emphasis on words such as "interface," "reliability," and "adaptation" in describing or explaining some of the new technologies. As a result of this phenomenon, organizations that pursue single objectives may be less suited for survival than those that consider a broader range of issues that optimize the human, organizational, and technological elements.

At the institutional level, private enterprises are the principal instruments in many countries for developing and using technology, although governments play an important enabling role. The task of private enterprises is to be knowledgeable about the current state of science and technology, to understand the needs of the marketplace, and then to create technologies, products, and services that best meet those market needs. Morris Tanenbaum pointed out that this endeavor embraces many disciplines (basic science, engineering, production, distribution, marketing, and finance) and individual motivations. Many participants and observers of the contemporary technological scene propose that we are going through a period of discontinuous change as the breadth of technological applications expands and the time scale of change becomes shorter. For example, markets are becoming more global as transportation and communication speed the flow of knowledge of new products, and greater investment is being made in research and development (R&D) as technological capability has expanded. This process has placed new demands on organizations as they strive to obtain quick and effective market information and access, recoup their R&D investment more quickly, and recognize the importance of sharing technological capabilities. This is particularly true with regard to the information technologies—the one technology most rapidly changing other technologies. It achieves its greatest power when it is most global; where it provides the means to obtain access to the information systems of other countries and establish arrangements that promote the transfer of technology.

Government plays a central role in technology issues at the national level. Technology has now become a part of almost every political discussion as politicians have realized the impact of technology on world events. Governments vary in the way they influence and exploit technological changes, for example, through regulation, procurement, protectionist policies, and support of R&D. Public attitudes among various countries also differ, and these differences can affect governmental technology policy. "Given the fact that there is no 'correct' way of dealing with technologies which is applicable to all countries," Sir Robin Nicholson commented, "each country must find its optimum way depending on its history, institutions, and public attitudes." This implies that countries will move forward at different speeds, creating imbalances among nations. In this respect, multinational corporations, responsibly managed and sensibly treated by the countries in which they invest, and transnational joint ventures serve an important function by promoting global equilibrium.

From an international perspective, the main issue is to sustain and improve world growth and improve growth per capita. This breaks down into the problems of Western Europe, Japan, the United States, Eastern Europe and the Soviet Union, and the problems of the more and less advanced developing countries. Robert Malpas noted that it becomes essential for all these players to harness technology for growth; however, this effort is frequently constrained by protectionism, concerns about intellectual property, the demands of international marketing and finance, and, of course, national security. The net result appears to be that emerging nations, with a few exceptions, have even more difficulty achieving the growth necessary to close the gap with leading nations. Among the trends at the international level that can help sustain and improve world growth: the rebirth of interest in manufacturing, the spread of expert systems which multiply skills and help in the industrialization process, the acceptance of multinational corporations, the privatization of various industries, and the increased interest of governments in technology.

As evidenced by the papers in this volume, these four relationships at the human, institutional, national, and international levels permeate discussions on the globalization of technology. In his keynote paper, Simon Ramo maintains that technological issues lie at the heart of most of the social, economic, and political issues of today, sometimes causing problems but more often offering possibilities for their solution. From this perspective, Ramo goes on to make several intriguing predictions about the role of technology in the future. Particularly powerful influences on the diffusion of new technological processes and products will be governments, corporations, national security concerns, and the rate of advances in scientific research. Technological discovery will become a global rather than an individual or national endeavor. As a result, new mechanisms will be developed to facilitate the flow of technology, despite protectionist-nationalist tendencies to stem the free exchange of information. One of these influences impeding the flow of technology is national security concerns. Ramo, however, is optimistic about the direction of the two superpowers, predicting that offensive forces will be reduced, thereby lessening interference with the flow of advanced technology and allowing the application of military technologies to peacetime applications in manufacturing, transportation, and services.

In scientific research, Ramo reiterates his belief that the expense of conducting such research, particularly in "big science" areas such as super colliders or in outer space, and the recognition that such knowledge must be shared to achieve maximum progress are driving scientists toward international cooperation. Since the role of government in setting a national direction for technology is so pervasive, its relationship to the private sector in the productive use of technology will continue to be problematic. Yet, Ramo argues, it is only the government that can perform the regulatory functions necessary for the smooth operation of free enterprise activity that makes use of new technologies. It is also the government, he says, that will be the primary obstacle to diffusion of the benefits of technology to world society.

As experts on the costs and benefits of developing technology, engineers are in a key position to contribute to policy formation of these issues. For engineers to better prepare themselves for the future, Ramo suggests that engineering education place more emphasis on the links between engineering and its societal applications. The result, he says, will be engineers equipped to play a broader role in influencing government policies and practices regarding technological advance.

Umberto Colombo's analysis of technological and global economic issues emphasizes the impact of the technological revolution on production methods, types of products, labor markets, and on the importance of manufacturing to the economy. He compares manufacturing to agriculture—although it will no longer dominate the economy or provide the majority of jobs, it will continue to perform an important function even in a service-oriented society. Certain key technologies are bringing about this transition, both creating new industries and rejuvenating mature ones, and in the process are changing patterns of development throughout the world. The rapid spread of innovation makes it imperative that firms quickly exploit any competitive advantage. Moreover, their increased ability to operate in the global marketplace reinforces the importance of cooperative agreements to advance innovation. Another force driving the trend toward cooperation is the increasingly scientific nature of technology, which requires that firms take a cross-disciplinary approach to solving problems. Colombo also argues that the technological revolution brings about a "dematerialization" of society, one element of which is that fewer raw materials are now needed to achieve a particular level of economic output and income generation.

The globalization of technology is being spearheaded by North America, Western Europe, and Japan. Despite their influence in shaping a new pattern of global competition, each has unique problems. The United States, though a leader in developing emergent technologies, is facing the double threat of enormous budget and trade deficits as well as deindustrialization of traditional economic sectors. Japan, which has demonstrated enormous success in commercializing new technologies, has an economy excessively dependent on exports. Western Europe has the cultural tradition and core of excellent research groups to facilitate its leadership in the technology arena, yet it lacks the cohesion necessary to develop strategic initiatives in important sectors.

Colombo optimistically concludes that globalization will bring the emergence of many small and medium-size multinational firms that will rely on a network of technology alliances. Governments will provide oversight and strategic direction. The impact on developing countries will be enormous. With the help of new technologies, Third World countries can transform their raw materials and energy into value-added commodities and thereby accelerate economic development without dysfunctional effects. It is the responsibility of developed countries, Colombo concludes, to see that this happens.

Though desirable, the alliances proposed by Colombo are not easily established. As Gerald Dinneen points out in his paper on trends in international technological cooperation, international arrangements, whether they be international marketing organizations, joint ventures, or creation of subsidiaries, are necessary if industries are to get a proper return on investment and remain competitive. However, the "not-invented-here" syndrome, differences in standards, lack of protocols for transmission of data, and especially protectionist sentiment prevent companies and countries from collaborating. Despite these barriers, Dinneen says, international labs and exchanges of scholars and students in schools of engineering have been effective mechanisms for fostering international cooperation.

Presenting the European perspective on technological cooperation, Harry Beckers comments on the impacts of the dissimilarities in the ways academicians and business people conduct research as well as differences in R&D support in the United States, Western Europe, and Japan. Western Europe, he says, faces the unique difficulties posed by its diversity and nationalistic tendencies. Nevertheless, there are a number of EEC programs that facilitate international cooperation among various countries, thereby helping to bring about "Europeanization" in the technology sector.

Papers on three of today's most crucial technologies—software, materials science, and information technologies—illustrate how the nature of the technologies themselves has created a global environment for research and applications despite the barriers mentioned above. George Pake describes a number of key advances in software: architecture of hardware systems used for software development; advances in writing, editing, running, and debugging of software; development of different programming languages; and systematic forward planning and task analysis. The creativity so evident in software technology today is not in danger, Pake says, despite the trend toward greater standardization and the possibility that ossification of the development system could occur in the future.

Pierre Aigrain addresses several provocative questions about materials, particularly pertaining to the rate at which discoveries are made, the extent to which applications are found, and the impact of these discoveries on industry and society. Citing the influence of the market and the continued interaction between science and materials research, Aigrain predicts that the rapid trajectory of materials discovery will continue. However, processing costs, rather than the costs of the materials themselves, prevent materials from widespread application. The development of superconductors illustrates this point, and he concludes with a description of the impact these new materials in particular will have on industry and society.

Lars Ramqvist provides insight on several of the cutting edge technologies that have had a major impact on information technologies. These include VLSI technology, computers, software and artificial intelligence, fiber optics, networks, and standards. In addition, he looks at three main applications of information technologies—normal voice telephony, mobile telephony, and data communications—assessing, first, the current state of the art and, second, projections for the future. Ramqvist concludes that because information technologies allow for the dissemination of information, and thus understanding, they will form the basis for a more equitable, humane society.

Hiroshi Inose examines the telecommunications sector from a different angle—the effect of globalization on the entire industry. Particular technological advances, for example, the convergence of service modes and the microelectronics revolution, provide economies of scale but also require rapid inputs for capital investment. Among the problems and challenges Inose addresses are the software crisis, or the high cost of developing more sophisticated and diversified software; structural changes in industry, particularly in job design and labor requirements; standardization and maintaining interoperability between systems and equipment; reliability and security of systems against both external and internal disturbances; and integrity of information and protection of privacy. Like Ramqvist, Inose views telecommunications technology as the means to promote mutual understanding and cultural enrichment worldwide.

Perspectives on the impact of technology on another industrial sector—construction—are presented by Alden Yates who describes the most significant trends in the areas of construction-related design, construction equipment and methods, automation and expert systems, and construction management. Computer-aided design has, among other things, improved communication between designer and supplier and speeded up the design development process. Increases in productivity are being achieved through off-site fabrication and assembly and robotics. Logistics practices, skill requirements, and labor-management relations are also changing as a result of these new technologies. Yates suggests that improved management methods and automation hold the greatest potential benefit for the construction sector, and that to remain competitive in the global marketplace, firms must look at their R&D commitments. In the long run, however, the effectiveness of management will determine success.

Pehr Gyllenhammar makes a complementary point about the importance of management practices in his paper on the manufacturing industry. To claims that the manufacturing sector is on the decline in an increasingly services-based, information society, Gyllenhammar responds that the manufacturing industry is adapting to today's environment. One of the most influential changes has been the new technologies employed in the automotive sector, including new engineering materials, computer-aided design, robots, and microcomputers. These new technologies mean that decision making can become decentralized and that small-scale manufacturing can be cost-effective. Another important factor changing the manufacturing industry has been new demands from employees and customers, what Gyllenhammar refers to as the invisible contract between them and the corporation. In fact, the new technologies have brought about important changes in the way work is organized. Less desirable tasks have been taken over by robots; light, flexible technologies allow workers to organize themselves so that they command the technology instead of vice versa; and new materials-handling mechanisms permit the layout of equipment to fit particular work organizations. The challenge for managers lies in organizing production so that they can develop their workers through both technical and leadership training. To accomplish this goal, it will be necessary for the manufacturing industry to take a longer term perspective and use "patient capital" rather than striving for a quick return on investment. Gyllenhammar concludes that a viable manufacturing industry is necessary but not sufficient to solve the problems of unemployment and slow growth.

The manufacturing industry is also the subject of the paper by Emilio Carrillo Gamboa; however, he discusses the issue of production sharing as both a result and a means of globalizing industry. By moving production facilities abroad to low-wage developing countries, firms manufacturing products that have entered the downside of the product cycle can maintain a competitive cost advantage. Mexico, in particular, has become an important production-sharing partner for the United States because of proximity, demographic factors, and the Mexican economic crisis which has resulted in lower wage levels that are competitive with labor costs in the developing countries of Asia and government programs that support production-sharing.

The *maquiladoras*, or production sharing sites, have been the subject of debate in Mexico for a number of reasons: the benefits of foreign-owned assembly services are not extended to the rest of the economy, the *maquiladoras* do not absorb traditional unemployment, and they are too vulnerable to swings in the U.S. economy. In addition, some of the plants have been criticized for their poor working conditions. Nevertheless, the author contends that they are an important source of income, employment, and foreign exchange, and proposes that the production sharing offers significant economic opportunities if the competitive advantages of Mexico as a production-sharing site are improved and assembly activities are more closely linked with the domestic economy. Carrillo Gamboa acknowledges the objections to offshore production sharing but suggests that its economic and political advantages far outweigh the disadvantages.

Further discussion of specific regional issues concerning technology's impact on development is provided in papers by Jan Kolm (Pacific Rim), Enrique Martín del Campo (Latin America), and Ralph Landau and Nathan Rosenberg (United States). In his paper on the consequences of globalizing industry in the Pacific Rim, Kolm uses a theoretical construct based on the technological complexity of goods and the product cycle to describe some general trends in the region's economic development. For example, gross national product (GNP) has increased rapidly due to the globalization of industry, and export-driven economies have helped the Pacific Rim nations overcome the disadvantages of scale and the shortage of foreign exchange. Kolm asserts that progress in the region is likely to continue, considering that there are suitable gradations of development, ample raw materials in the region as a whole, and a populace that has demonstrated its ability to cope with technological change. The focus of the paper then narrows to an examination of the problems and challenges facing the major groupings of Pacific Rim countries: the Association of Southeast Asian Nations (ASEAN); the newly industrializing countries, in particular, the Republic of Korea; Australia; and the United States and Japan. Despite their diversity and the impediments they have faced in their industrialization, Kolm contends that technology transfer has been less problematic in the Pacific Rim than in other countries of the world, a sign of hope that competition can coexist with cooperation.

Enrique Martín del Campo deals specifically with the influence of technology on development in the Latin American and Caribbean countries. Shifts in economic strength and investment patterns influence the developing countries and make it imperative for them to develop strategies for growth through improved technological and entrepreneurial activity. Martín del Campo suggests that the region's technology strategy must combine development of both advanced and intermediate technologies, linkage of smaller and large enterprises, and diffusion of technological development through many sectors.

Because the economies of the region, like most developing countries, participate in the international sphere through foreign trade, competitiveness in foreign markets is crucial. The rate of innovation, the ability to apply advanced technology, the degree of capital investment, use of natural resources, and the existence of technological support services all affect the competitiveness of Latin America in foreign markets. Two major factors, however, hamper economic growth in Latin America. First, investment is curbed by an economically depressed environment, and second, global demand for the region's traditional exports is weak. Any plan to remedy these problems would require a strong technological component, including development of local capabilities in technology, internal and external transfer of technology, strategic projects that integrate science and technology, and government policies that support scientific and technological endeavors.

Ralph Landau and Nathan Rosenberg review the impact of technological change on U.S. economic growth. They cite several key influences on such growth, including technological innovation, high capital investment rates, and increased training of the total work force. The authors conclude that U.S. economic policies are not conducive to innovation and capital formation, and they propose strategies to ensure continued economic growth.

One change that poses both opportunities and difficulties is the rapid diffusion of technology to other countries. As a result, the exploitation of new technologies is no longer an exclusive strength of the United States. The maintenance of a high-wage economy will depend on the ability of U.S. firms to compete in international markets, particularly in manufacturing because of that sector's contribution to GNP, foreign trade, and national security; its purchases of services; and its productivity increases and consequent contribution to the overall economy.

Landau and Rosenberg also focus on the role of government in creating a favorable environment for business decision making. Policies that encourage personal savings from which investments could be made, reduce the budget and trade deficits, and support a long-term financial climate are essential. However, because U.S. business interests and government do not work as closely as they do in some other countries, Japan, for example, this goal may be difficult to achieve. If the United States is to remain competitive, increases in productivity must be sustained, and this will happen only if training, management, and investment in manufacturing and services are part of public and private strategies.

In the volume's final paper, Hajime Karatsu reminds us of some fundamental points about the role of technology in improving the quality of life. Technology is instrumental to economic growth, and as a result, economic strength is no longer a function of a nation's size and population, as it was before the industrial revolution. Although some will argue that technology is the cause of the problems resulting from industrialization, Karatsu describes how technology has been used to provide solutions to some crucial problems—the oil crisis and pollution—in his own country, Japan. Although not explored in his paper, one issue Karatsu's point raises is the extent to which technological decisions operate in concert with other strategic policies. For example, Japan's pollution problem, and that of many other industrialized countries, has been solved in part by the export of the pollution-causing industries to other nations.

Karatsu supports his views on the importance of technology to economic growth by commenting that Japan's methods of applying technologies have allowed it to achieve a 1986 GNP of $2.3 trillion, or 11 percent of the world's economic activity. One characteristic of Japanese methodology is that new, advanced technologies are applied in practical and simple ways that can be easily commercialized. He cites as an example the use of carbon fiber in golf clubs and fishing rods. This practice contrasts with that of the United States, where advanced technologies are frequently applied to complex products in the defense industry. A second aspect of Japanese practices in commercializing new technologies is their attention to incremental changes and improvements in product and process. Karatsu concludes by stressing the importance of technological cooperation so that standards of living can be improved worldwide.

The papers in this volume reflect a diversity of national perspectives on the impact of cutting-edge technologies on the individual, industry, and society; appropriate means for harnessing technology to facilitate economic growth for all nations; and the roles that should be played by institutions and governments in the emerging global economy. Nevertheless, agreement on several key issues is apparent: First, technology will continue to fuel economic growth and rising standards of living around the world. Indeed, technology's influence is pervasive, for it shapes trade patterns and policies, employment, and even relations among nations. A second area of consensus centers on the important role to be played by the engineering community in facilitating international technological advancement. As mentioned by Stephen Bechtel in his introduction of the keynote speaker at the convocation, ". . . we (engineers and technologists) can only benefit by being more attuned to the factors that influence each country's technological interests and capacities." Although this process is frequently constrained by national competitiveness concerns, Bechtel asserts that it is only through increased cooperation that nations remain competitive. "A nation's strength as a participant in the world economy is derived in part from its ability to adjust to rapidly fluctuating economic conditions and technological change. Cooperation provides access to regional and national trends in technology, thereby benefiting individual nations as well as the international engineering endeavor." Indeed, industrial competition can be a source of creative tension for the world economy when viewed within the larger global framework of cooperation directed at improving the quality of life for all.

Globalization of Industry and Implications for the Future

SIMON RAMO

THE KEY PROBLEMS OF SOCIETY TODAY are rarely categorized by governments as problems of science and technology. They are assumed to be social-economic-political problems, such as avoiding war, building the economy, curbing terrorism, fighting poverty and disease, or preserving a healthful environment. We can predict, however, that it will become clear that all social-economic-political issues intersect and that issues of technological advance are right in the middle of every intersection, sometimes causing the problems, more often offering possibilities for their solutions, and frequently providing opportunities for the world society to rise to new, higher levels of productivity, satisfaction, and happiness.

Of course, trying to predict the future with perfection is not defensible. But if we are active in some field, we are bound to be aware of important trends in it, and we should take our anticipations of significant future events seriously. We should force ourselves to list potential developments regularly, estimating for each the probability of its occurrence and the importance of the event should it occur. Then, for those happenings we consider both highly probable and significant, we should ask ourselves what we can do early to enhance the positive and suppress the negative consequences.

At present we see that the impact of advancing technology on the social-economic-political framework of the entire globe is growing rapidly, yet its implications are far from being adequately explored. Moreover, the effects of technological change are underestimated as short-range; crisis-dominated problems monopolize the attention of most of the world's leaders. A severe mismatch is developing between accelerating technological advance and lagging social progress.

Thus, advances in information technology provide computer systems that can alter productivity and employment patterns to a much greater degree than the passing of new minimum wage laws can. Broadened global communications and transportation, resulting from technological breakthroughs, can speed up and link the world's operations far more than trade pacts can. Not only has nuclear weapons technology radically altered the dangers of war, but it dominates negotiations intended to prevent war between the superpowers, and the very awesomeness inherent in the technology has averted such a war, despite the noticeable and continual shortcomings of the political leadership involved.

Technological advance is becoming globally pervasive, and this leads us to another prediction. It is that the totality of advances produced by the international fraternity of nations henceforth will greatly transcend the new technologies generated by any one nation. To be sure, an individual engineer, scientist, corporation, or country may happen upon a great discovery or invention or may successfully focus resources for progress and attain leadership in one area. That entity might then possess an initial superiority—on that one item—exceeding the expertise or output of all the rest of the world put together. But if it is an important advance, then pockets of similar or greater concentration will pop up in many other places around the world almost immediately, and the total will soon dwarf the continued contributions by the source of the breakthrough.

The rapidly growing disparity between the total output of technology from all nations and the contribution of any single nation means that no one country in the future will be strong enough in new science and technology to depend solely on its own intellectual and physical resources to fight the competitive world battle. The prediction, more specifically, is that the effort to achieve technological advances will become so widespread, and engineers and resources to back them up so widely available in the world, that what happens technologically on the outside will become too important for any country to ignore, and a failure on its part to acquire and use external advances will be too penalizing to be tolerated.

Owing to the clear economic potential of technological developments, should we predict that national governments will universally reach agreements in the future for all technological advances to be made available freely to all nations, all peoples, all private entities throughout the world, no matter where the advances originate? Will all new technology be owned by no one exclusively and hence equally by everyone? Will inventions, information, and ideas be breathed in by every group, shared like the atmosphere that surrounds us all? Not quite. There is the certainty of entirely opposite government action, deliberate steps taken to impede the flow of technological advances, policies set up to seek a perceived national advantage. There is also the certainty that private entities in the non-Communist world will continue to have an interest in receiving *quid pro quos* for accesses to their technologies.

To probe for sensible predictions here, let us note some aspects of world trade policies. They will provide useful analogies.

Every nation has social and economic problems and a citizenry that looks to its national government to solve them. The movement of assets across borders is bound to affect conditions within each nation. Therefore, it is impossible for governments to keep their hands off this flow as they seek to give their domestic industry competitive advantages, to protect jobs for their nationals, or to exact revenues in return for privileges to export products into that nation or, in some nations, as the politicians try to curry favor with their constituencies. Accordingly, we can predict that the protectionist-nationalist approach to international trade will remain strong forever.

We should not predict a world totally in the grip of protectionism-nationalism, however, because the very different free trade alternative offers such powerful economic benefits. In this one-world approach, raw materials, manufactured products, services, money, management know-how, and even labor cross borders without constraint. Each nation offers what it has or can most sensibly produce to others at competitive free market prices. It acquires from other nations in turn what they can best present to the unfettered world marketplace. When each entity concentrates on what it is most suited to supply, those fortunate in their possession of natural resources, skills, and developed infrastructures admittedly enjoy advantages. But if the output of any country, richly endowed or not, is available to others in unrestricted return for whatever they can most readily contribute, then all tend to be significantly better off economically. Because of this, it is difficult to push free trade mechanisms aside entirely. Without doubt, that approach is also here to stay. The international trade of the world, it can be expected, will exist as a hybrid of both protectionism–nationalism and free trade.

Such a hybrid pattern, we can predict, will apply also to the flow of technology. Wherever it might first occur, every technological advance will tend to move to all those nations and private entities that want it, organize to obtain it, and are willing and able to pay or trade for it. Despite the permanent, simultaneous presence of forces working to restrict the flow, technological advance will come to be seen as predominantly a global, not local, phenomenon of generation, dissemination, and use. All nations, it can be predicted, will adjust their policies and operations to accord with this concept. Of course, the recipient groups will all have to be technologically advanced enough to be able to assess, select, and use effectively the world's generated advanced technology. In time, we can expect such competence to become universal.

A powerful influence militating today toward a higher rate of technology flow about the world is that of the private sector, the private corporations, the organizations whose objective is to achieve a satisfactory return on the risk investments they make to spur technological advance and exploit it. How will the strategies of such companies affect the global impact of technological advance in the future? We can predict that the private corporate world will become an ever-stronger force for dissemination of advanced technology, causing dissemination to be faster, more complete, and ubiquitous. The full use in the world market of every major advance a company develops will be regarded as mandatory for two reasons: (1) the return on investment can be greatly magnified by wide geographical dispersion of efforts to exploit the technology, and (2) failure to take a product worldwide increases the danger of losing out to competitors that do operate everywhere. Both factors will grow as the globe becomes an array of centers of technological strength, each center a candidate for creating, buying, leasing, sharing, and using advances in technology.

With the rapid advance of technology becoming a worldwide phenomenon, no one company, not even the largest, can hope to originate more than a small fraction of the evolving technology that will be key to preserving its position. Every company will recognize the growing danger that a novel product invention or manufacturing process, a newly developed material critical to performance or fabrication, or a creative application of recent science may make its technology obsolete and drastically change the economics underlying its endeavors. This will cause a great rise in efforts to buy and trade advances in technology.

As a result, we can predict that we shall see the creation of a major new kind of technological industry. A new class of multinational companies will take root and grow quickly—but not to make and sell a device or system. Instead, their product offerings will be technology itself, but not technology they originate. They will gather advanced technologies from sources all over the world and offer them to the technological industry everywhere. As agents of transfer they will be paid well for their services because of the values of the trades both to the originators and to the appliers of the advances.

It may become rare in the future for a typical company to contemplate keeping its technological advances long for its exclusive use. That company will expect that an avalanche of alternative developments will soon build enormous competition. At the same time, given a proper market system for technology trading, the more and sooner its novel developments are applied globally by everyone, the greater may be the returns to the developer. On the other side, a single company may find it costly to search the earth to locate the technology it should endeavor to secure and use, and then arrange to acquire access to it. The new technology-trading companies will help by setting up a fast and cost-effective technology exchange market.

In the future, when analysts study a corporation, trying to measure its present condition and future potential, we can predict they will add one important new item for study to their conventional examination of balance sheets, profit and loss statements, market growth rates for the products, and strength against competition. They will check also on whether that company is obtaining the technological advances of others economically and quickly and is employing them effectively.

We must now turn to a powerful influence that will restrict rather than enhance world technology flow—namely, national security or military technology. Unfortunately, we have to include among our predictions that the future, like the past, will not be characterized by universal, permanent peace. Technological advance will continue to be a key factor in military strength. Moreover, a sizeable fraction of the technological resources of the world will be committed to the development of weapons systems. Finally, advances in military technology will continue to produce peacetime spinoffs that will continue to affect the total world rate of generation of commercially useful new technology. All in all, global technological advances and their dissemination and use will continue to be affected greatly by technology's relationship to national security. What can we predict about that relationship in the future?

Consider first the most dangerous of military technology—strategic nuclear weaponry. Here we shall present an optimistic prediction. The involved nations of the world have together spent trillions of today's dollars—from the first atomic bomb research to the creation and maintenance of manned bombers, intercontinental ballistic missiles, and submarine-launched missiles. Such a massive commitment of technological resources has profoundly affected the global economy. If nothing changes, another trillion dollars will be expended during the rest of the century to ensure continued mutual deterrence. But we can predict that so enormous a future investment in further expansion of offensive nuclear weapons will not take place, because the weapons will be seen as not useful except to deter others from using theirs and, being useless, not affordable at such high economic costs.

Neither the United States nor the Soviet Union, we can predict, will launch a nuclear strike against the other, because the leadership in each nation will know that it would fail. To be successful, the first striker's blow would have to leave the stricken nation with no significant capability to retaliate, and the aggressor would have to be assured that it would not suffer serious injury from unavoidable nuclear after effects. Neither result is possible. Even with a 99 percent score against the other nation's retaliatory force, the remaining 1 percent (hundreds of nuclear bombs) would make the potential consequence unacceptable to the striker, because today's weapons can be so destructive and the present inventories are so large. But no competent weapons engineer would expect anywhere near perfection from so complex an operation, one not completely rehearsable even once. Think of the timing problem. Considering that to launch one craft at a scheduled moment is a challenge, imagine coordinating thousands of launchings from thousands of square miles of land and sea so that all offensive warheads will arrive simultaneously. If the first-strike nuclear weapons were to reach their targets over a spread of, say, 30 minutes, the other side might release their unhit retaliatory missiles immediately after the first weapon has detonated. Most of the offensive bombs then would land on empty silos, the retaliatory missiles having been sent off to blast the first striker.

The United States has an alarming budget deficit ($200 billion in 1987), and the USSR has a critical need to find investment funds to attain economic growth. The United States is struggling to maintain its standard of living, the Soviet Union to get its raised. The powerful resulting economic pressures combined with the perceived uselessness and dangers of nuclear offensive weapons systems will drive the superpowers to agree to large-scale reductions, both becoming confident that they can fear, scare, distrust, and deter each other at a much lower level of expenditure.

This leads to the prediction of a stable future pattern for nuclear weaponry, one that will last for years, deter nuclear war, and be tolerable in cost. The reduction of offensive nuclear weapons down to, say, 10 percent of present forces is reasonable to anticipate. Verification systems can and will be worked out to ensure that such agreed-to levels of reductions take place and are maintained. Defense systems, practical against a ten-to-one reduced offense, then will be put in place by both sides at reasonable cost, with the capability of shooting down 90 percent of incoming missiles. The number of damaging bombs from a first strike surviving to arrive on target, then, would be a tenth of a tenth, or a hundredth, of the present forces. Launching an attack would become preposterous. The installed defense systems, moreover, would provide insurance against an accidental launch, cheating on arms reduction agreements, or a deliberate attack by a terrorist nation.

Even for European theater military forces, favorable predictions deserve to be taken seriously. Moreover, there are additional aspects to consider here, different from the strategic nuclear weapons situation, that will influence most particularly the way advancing technology will affect the nature and vigor of the global economy.

To see why, we start by predicting that the European military strategies of both the East and the West will soon come to be based on nonnuclear military force structures. One consequence will be that European theater military strength will cease to be rated mainly by firepower, numbers of soldiers, and numbers of tanks, airplanes, and other equipment. The true strength of the forces in the future will be increasingly manifested by sophisticated technology for command, communications, intelligence, and reconnaissance and for the launching and guidance of robotic air, ground, and sea weapons in a broad variety of forms. Proper application of advanced electronics and robotics can make a defense force greatly more capable than a larger offensive army less well equipped with such technology.

The basic technologies involved here happen to be very close to those that are needed for increasing productivity in the peacetime operations of the world. Technologies developed for the military, such as advanced computers, fast and powerful semiconductor chips, and versatile electromechanical devices for automatic control, will be close cousins to technologies useful for improving manufacturing of commercial products and for employing information for superior management of nonmilitary industry, transportation systems, and governmental and professional services. These associated nonmilitary developments will come to be seen as extremely important economically, will be sought avidly by all nations, and will become increasingly popular in world trade.

We can predict, then, that nations on both sides of disagreements and distrusts will tend to move toward negotiating reductions of offensive forces in the European theater. The greater the negotiated military reductions, the less the governments will feel required to interfere with the international flow of advanced technology and the more the nations of both the East and West will be able to invest in commercial application of the basic technologies and realize the economic rewards of the resulting trade. Trading of technological products will become freer between the North Atlantic Treaty Organization (NATO) and Warsaw Pact nations, even as they continue to regard each other as potential enemies.

Technological advances stem in part from scientific discoveries. Let us now shift the focus of our prediction to basic scientific research, where the objective is to push back the frontiers in understanding the laws of the physical universe. What changes do we see in the future patterns of scientific efforts? How might these changes affect, in turn, the way technological advances will affect the global economy?

We can see what is likely to happen by looking at two areas of scientific research. One is the building and application of colossal accelerators of the tiny particles that are the constituents of matter. The other is the exploration of outer space. Both are characterized by requirements for massive expenditures. The high costs have already encouraged some international cooperation. We can predict that greater cooperation will occur in the future because the expense of designing, building, and maintaining the needed equipment and facilities is becoming too high for any one nation.

The particle accelerators offer the promise of giant steps in understanding the makeup of matter, the relationship of matter to energy, the formation of the universe, and the forces that control the dynamics of all physical phenomena. In outer space, manned exploration of Mars exemplifies the challenge as well as the enormous costs and complexities that will drive us toward international cooperation.

Scientists worldwide agree not only on the importance of research but also on the idea that it is mandatory for the acquired data to be made available to all scientists in all countries. This view is held by scientists universally not only because of a high sense of social responsibility but also for the practical reason that participation by all the world's scientific brainpower is required for maximum progress. Eventually the necessary international approach to organization and sponsorship of large-scale frontier scientific research will act as a strong catalyst to the formation of friendly teams among nations in commercial technology.

In both national security matters and pure research into the laws of nature, the impacts of scientific discoveries and technological advances on the global economy and society tend to be dominated by decisions made by governments. To be sure, national academies, universities, private corporations, and occasionally even individuals of exceptional public visibility and stature help shape the decisions. But because governments provide the funds for huge research projects and weapons systems, in the end, the governments are the bosses.

In civilian commercial technology in a democracy, government is not in control but is indirectly forceful in the setting of priorities and pace. Government influence is comparable with the combined influence of free market forces and private decision makers, even though the latter determine the allocation of their privately held funds. When we probe the future effects of technological advance on the world's economy, what should we predict about the relationships between government and the private sector? We are way behind in studying this role-setting issue and we are not likely to catch up. Confusion is likely to reign in the future, greatly impairing attempts to use advancing technology fully for the benefit of society.

For example, consider information technology in the United States. Information is essential to human activities, and advancing technology is revolutionizing the way we acquire, store, process, ponder, transmit, and employ information, enabling us to perform these operations a thousand—sometimes a million—times more rapidly, cheaply, accurately, and effectively. If today's information technology were put to work fully in the United States wherever a substantial economic benefit would result, the cost of the installations would exceed a trillion dollars. An investment of that size and the resulting financial returns, even when spread over several years, obviously will exert a powerful effect on the economy. The social impact will be even greater, because virtually all tasks will be accomplished more efficiently and many jobs will be eliminated even as new ones are created. The greater the potential economic gain, the greater the pressure for speedy implementation. The quicker the changeovers, however, the rougher will be the societal dislocations and the more certain it will be that the government will be expected to manage the social alterations.

Consider one manifestation of new technology–society interactions resulting from advances in information technology. In the future, two-way national networks in the United States will come into being, connecting hundreds of millions of computer and communications terminals in offices, factories, homes, schools, hospitals, airlines, and almost everywhere else. The information flow and processing in the network will vastly exceed those of the present telephone, radio broadcast, cable TV, and computer networks. All sorts of information will move all about the nation and will be integral to the nation's functions. Thousands of private firms will be involved in producing the equipment, maintaining the networks, furnishing the information, and designing and setting up interconnections between the networks and the users. Hundreds of millions of hardware boxes will be used.

Free enterprise activity will burgeon, but government involvement will be inevitable and essential. Who but the government could create and regulate the necessary standards for the electronic expression of information throughout the networks and for the hardware and software to keep the whole national complex from developing into a chaos of digital data? Who else could protect privacy yet guarantee free access to the networks that will enter virtually all homes and workplaces? Who else could prevent monopolies of information supply, transmission systems, and equipment manufacture? Who but the government could police to prevent malicious input of misinformation or tinkering with data, or fight fraud and set rules as money is shifted electronically and as distinct private activities are linked to attain efficiency in production and distribution?

Government action is critical in all these functions, but the U.S. government is not now organized for it. Arranging for the government's proper role will not be easy. If the government performs well the tasks it alone must handle, then the private sector will do a superior job of introducing new technology. Investors will see a rational, stable environment for long-range development if information technology and the free market will work to encourage investments. If the government defaults, it will be the limiting factor in the use of new information technology.

Applying information technology fully will create new policy problems not only domestically but also internationally. Just as each nation will have to regulate electronic information flow within its borders, it will also feel compelled to control the flow across those borders. Although in some ways information is like a product in international trade, in other ways the transit of information between nations exceeds in importance the movement of goods. For instance, the advent of electronic money (fund transfers and commitments made virtually instantaneously by electrical signals) will force every nation intent on administering its money supply and its banking operations to monitor the transmission of financial data. Or consider that at some future time, when much of production and distribution will be computer automated, the flow in information networks within a nation will constitute the very heart blood of its economic life. Governments hence will want to protect that flow. But manufacturing in one country typically will depend increasingly on the timely delivery of material and components ordered from another. Yet, to interfere with information flow tying production operations together internationally will harm economic efficiency. Will governments allow free border crossing of goods whose shipments are automatically scheduled by interconnected information networks? How will they set international rules to control the interconnections and the shipments?

The information network is but an example of how the full employment of new technologies will require the defining of new roles for governments domestically and new challenges for international agreements. This is a terribly complicated process because governments are bureaucratic and politically controlled. In general, only a crisis receives the direct attention of leadership, whereas systems and long-range problems that involve integration of many factors and require new patterns of organization are very difficult for governments to handle. We can predict that whether the advanced technology be in the field of information, energy, environment, transportation, genetic engineering, oceanography, weather control, or whatever else, governmental response will lag behind the technological advance. Governmental planning, evolution, and implementation of policy always may be so late, we must assume, that government effort will be concentrated entirely on the disbenefits surfacing from technological advance, and only when those disbenefits are perceived as causing a serious political problem. We can predict that government will be the bottleneck in determining the extent to which the benefits of advanced technology are realized by the world society.

Recall that we engaged in this exercise in prediction with the idea of noting potential developments and then asking what we might do ahead of time to accentuate the positives and eliminate the negatives. The engineers of the world are the experts on the nature of developing technology and its costs, but not on its societal impacts. They are best able to estimate the speed of development and how speed and quality of achievement will relate to the allocation of resources. If the future will be characterized by the increasing impact of technological advance on the global economy, then the world's engineering leadership needs to be more interested in this impact in the future than it was in the past. Engineering leadership should be consulted more and contribute more to policy formation.

When we say engineering leadership we are describing the world's academies of engineering. What should the academies be doing in sensible anticipation of problems, dilemmas, and opportunities for the world ahead? What can the academies do to ensure the maximum contributions from the world fraternity of engineers?

For one thing, the academies should help everyone to understand what engineering is really supposed to be. Probably all the dictionaries in all the languages of the world use the same short definition: Engineering is the application of science and technology to provide benefits for the society. In citing this definition we are not advocating that every engineer should be an expert in science and technology and also be a sociologist, economist, and political scientist, although it would probably be advantageous if there were more such hybrids. It is rather that the engineering profession and the general public should have the image of the profession contained in the definition. Even if an individual engineer is best employed in designing a microsemiconductor chip, an automobile, an airplane, or a petrochemical plant, that engineer should think of his or her profession as being broader than his or her duties.

Not only the academies but virtually all other professional engineering societies, as well as the engineering departments of our universities, have tended in the past to be preoccupied with the science and technology basic to the engineering profession. The academies should encourage universities to introduce students—not just engineering students but all of them—to the idea that engineering not only encompasses the science and technology basic to engineering accomplishments but also links technology to its societal applications.

If the image of engineering is properly broadened, disseminated, and accepted, then the leadership of the engineering profession surely will come to play a much larger part in the society's decision making as it arranges the way technological advances will alter the global economy and civilization in general. Much more often than they do today, engineering leaders will find themselves interested in adding the political dimension to their careers, and more will become leaders in government.

We should not realistically expect engineers to compete numerically with lawyers as members of the legislative bodies of the world. The medical profession probably furnishes a better example. Its leadership expects to be involved and listened to seriously when the government sets rules and regulations pertaining to the allocation of resources to protect the public's health and to advance the field of medicine. Engineering leadership exerts much less influence in matching technology to societal needs. Engineers do mainly the technical work called for by the policy and priority decision-making process. The future must be different, and we can predict that it will be.

The Technology Revolution and the Restructuring of the Global Economy

UMBERTO COLOMBO

THE WORLD IS IN THE THROES OF A TECHNOLOGICAL REVOLUTION that differs from the periodic waves of technical change that have marked the progress of industrial society since its origins 200 years ago. A shift is occurring in the sociotechnological paradigm that underlies our current sophisticated industrial structure. This old paradigm consists of the mass production of essentially standardized goods in ever-larger units; an emphasis on quantitative goals for production, requiring ever higher inputs of capital, energy, and raw materials to produce more and more; and little attention to environmental impact, resource use, and conservation issues. In contrast, the new paradigm taking shape is identified with an emphasis on quality and diversification of products and processes, diffusion of small but highly productive units that rely on new technologies and are linked to a process of decentralization of production, adoption of process and product choices requiring far less energy and materials input per unit of output, and a greater awareness of the need to preserve the quality of local and global environments.

Thus, we are in a period of transition between two epochs, a time comparable to the industrial revolution, when the steam engine was introduced and coal was the emerging energy source. Then, as now, there was widespread fear of the future, a fear derived from the difficulty of even imagining the range of opportunities that an ongoing revolution brings in terms of new activities and related jobs.

During a transition of this magnitude, past equilibria are disrupted and conditions of mismatch occur in labor markets. The demand for new jobs and skills increases, and old activities disappear or lose their importance in the marketplace. These changes are visible; their impact is almost immediate. It is now clear that the paper-free office is going to be widespread in a few decades, and in fact, we can see its beginnings with increased office automation, the spread of word processors, and the adoption of integrated workstations. The human-free factory is also in sight. With increasing automation and robotization, it is not only blue-collar jobs that will be eliminated. The change is more profound. We are witnessing the sharpened decline of the factory as the primary function and chief labor-absorber in industry. Research and development (R&D), marketing, finance, corporate strategy, legal affairs—functions that previously were to a certain extent ancillary to production—are assuming the center of the stage. Now manufacturing itself becomes ancillary and often even a candidate for contracting out.

This does not mean, however, that manufacturing technologies are becoming secondary in importance. The contrary is true, and here, too, history offers a parallel. Today's situation presents an analogy with the position of agriculture after the industrial revolution. All through the history of industrial society, agriculture improved its output and productivity enormously, although it no longer dominated the economy and was not the main source of jobs as it once had been. Industry will repeat this pattern, as the transition to a postindustrial, service-oriented society is completed.

The present era of change is being brought about by a whole cluster of technologies, some of which have an exceptional capacity for horizontal diffusion in all sectors of the economy and society and an equally exceptional capacity for cross-fertilization. Key technologies in this category include the microelectronics–information technologies complex, the biotechnologies, and the new materials science.

This process of technological change spurs structural changes in the economy and society. Mature sectors (such as machine tools and textiles) can be rejuvenated by grafting new technologies onto their processes and products. When this rejuvenation occurs in industrialized countries, these traditional sectors take the lead in international competition. Italy is a case in point, since Italian prosperity is in no small measure due to the restored competitiveness of such sectors. These sectors demonstrate a highly flexible approach to production, making possible less standardized products specifically designed to satisfy the tastes and needs of customers. They also demonstrate considerable creativity through attention to design factors and closer links to the market and its fluctuations, attentiveness to moods and fashions with highly imaginative marketing, and a capacity to absorb new technology and indeed to interact with it to generate improvements and adaptations.

The fact that in Italy these sectors tend to consist of dynamic, small- to medium-size firms organized in industrial districts is extremely important. Such districts operate as coalitions of competitors, interdependent yet united by a common goal. This pattern encourages the diffusion of technology through all firms in the district. This is in marked contrast to experience elsewhere when competing firms tend to keep technological advances closely to themselves in the hope of retaining competitive advantage. Ideally, rejuvenation of mature sectors is a "bottom up" process, though in Italy, for example, the European Nuclear Energy Agency offers a significant "top down" contribution in terms of information, expertise, support research and development, and project management.

Mature sectors that undergo such technological renewal and then strive continually to keep abreast of technological developments and market trends can retain competitiveness even in the face of increasing international competition. This pattern is one of the elements suggesting that long-established concepts of comparative advantage and ensuing international division of labor must be challenged. In today's new economic environment, the availability of abundant, low-cost raw materials and a pool of cheap labor is no longer enough to ensure market advantage to developing countries. But the emerging technologies are not the exclusive domain of advanced countries, and their intelligent application in developing countries may speed up their economic growth and open possibilities for decentralized patterns of development.

Until recently in the advanced countries, the main technological innovations in production have involved mass production and standardization. The emerging technologies make it possible to give an effective answer to the demand for diversification, product customization, and personalization. Thus, the structure of supply is becoming more flexible and innovative. In other words, it is now possible to combine small-scale production units with high productivity and high quality efficiently at increasingly accessible prices. We may therefore say that "small becomes beautiful again," although not in the sense that E. F. Schumacher used this phrase in the early 1970s.

The pace of innovation is extremely rapid. No individual firm or country can hope to gain or retain technological and market superiority in any given area for long. The pressure of competition and the rapid spread of production capabilities, innovative ideas, and new patterns of demand compel companies to measure themselves against rival firms at home and abroad early in the production cycle, and then rapidly exploit, in the widest possible market, any competitive advantages that arise from a lead in innovation.

We are witnessing a compression of the time scale by which new technology is introduced, with ever-shorter intervals between discovery and application. This compression is especially apparent in microelectronics and the information technologies, sectors in which international competition and academic and industrial research activities are intense. This phenomenon is widely visible though not universal. In some sectors (specifically, though not exclusively, those involving the life sciences) longer periods are imposed by the need for testing to satisfy regulatory criteria. Examples here come from the pharmaceutical and agrochemical industries.

Simultaneously, firms acquire more strategic space in which to operate. In the past, the smaller the firm, the narrower its natural geographic horizon.

Today it is possible for both large and small firms to think in global terms. This new perspective implies the need for all interests, large and small, to seek arrangements such as transnational mergers, joint venture agreements, consortia, and shared production and licensing agreements with other companies. The partners often bring complementary assets: investment capital, market shares in different geographic areas, technological capabilities in adjacent domains, and different strategic approaches to advance innovation. In this way returns in different countries can be maximized rapidly. This worldwide change is being spearheaded by the industrial democracies—the countries that possess major resources in science and technology, innovative capability, and investment capital.

Today's technology is becoming more and more scientific. Not only is it created and developed on scientific bases, but it also generates fundamental scientific knowledge. The discovery of new superconducting materials, for example, is simultaneously a great scientific achievement that implies fundamental advances in our understanding of the behavior of matter in the solid state and a technological invention that is immediately open to extraordinary applications in many fields, from energy transmission to computers and from high-field magnets to nuclear fusion. The development of artificial intelligence is another example of the increasingly scientific nature of technology; this effort requires the cooperation of the most disparate disciplines and in turn holds the potential for application in a wide variety of fields. These examples illustrate how the narrow, specialized, compartmentalized ways in which problems typically were approached in the past are giving way to a more global approach that breaks down the barriers of single disciplines to obtain a unified, cross-disciplinary vision.

Another unique aspect of the present technological revolution is that it brings about a dematerialization of society. In a sense, dematerialization is the logical outcome of an advanced economy in which material needs are substantially saturated. Throughout history there has been a direct correlation between increases in gross domestic product and consumption of raw materials and energy. This is no longer automatically the case. In today's advanced and affluent societies, each successive increment in per capita income is linked to an ever-smaller rise in quantities of raw materials and energy used. According to estimates by the International Monetary Fund, the amount of industrial raw materials needed for one unit of industrial production is now no more than two-fifths of what it was in 1900, and this decline is accelerating. Thus, Japan, for example, in 1984 consumed only 60 percent of the raw materials required for the same volume of industrial output in 1973.

The reason for this phenomenon is basically twofold. Increases in consumption tend to be concentrated on goods that have a high degree of value added, goods that contain a great deal of technology and design rather than raw materials, and nonmaterial goods such as tourism, leisure activities, and financial services. In addition, today's technology is developing products whose performance in fulfilling desired functions is reaching unprecedented levels. For example, it is now possible to invent new energy sources that have energy densities far exceeding those of raw materials. One kilogram of uranium can produce the same amount of energy as 13 U.S. tons of oil or 19 U.S. tons of coal, and in telecommunications 1 ton of copper wire can now be replaced by a mere 25 or so kilograms of fiberglass cable, which can be produced with only 5 percent of the energy needed to produce the copper wire it replaces. Decoupling of the amount of raw material needed for a given unit of economic output, income generation, and consumption of raw materials and energy is an essential element in the dematerialization process.

But present trends go beyond this. Dematerialization also includes the emergence of what has been called an "information society." The speed of information flow and its impact on the rate of innovation and diffusion and the capacity to overcome barriers have enormous implications.

World society is becoming more open; interdependence is increasing. World trade in goods and services has reached $3 trillion. This is certainly a high figure, but surprisingly, it is more than an order of magnitude lower than the volume of foreign currency transactions ($35 trillion) and of the estimated annual turnover of the London financial market alone ($75 trillion, or 25 times greater than the entire world's visible trade). This is part of what is increasingly being termed the globalization of business and finance.

The comparison between the various forms of trade and transactions is, however, a matter of concern. It might be an indication that conditions for profit increasingly are more favorable in financial speculation than in capital investment in a world that still greatly needs economic growth and opportunities for employment. The alarming indebtedness of developing countries and the massive transfer of resources to advanced economies in interest payments are another facet of this problem.

But globalization affects all sectors of the economy. As noted earlier, the present wave of innovation, technological and otherwise, is spearheaded by the industrial democracies: the countries of North America, Western Europe, and Japan. Kenichi Ohmae (1985) refers to this as the emergence of the "triad," and advocates a strategy of cross-cultural alliances in the industrial and business communities that will allow innovative companies from the three corners of the triad to become real powers, thus shaping a new pattern of global competition.

In this context, protectionism and defensive attitudes are losing bets. It is not by chance that even a superpower—the USSR—that had built barriers around itself and was striving to compete and advance by planning its economy in isolation is now being forced to come to terms with this new reality and open up to the opportunities afforded by technological change. The implications of Gorbachev's new course for the organization of Soviet society are immense, and the bureaucratic resistance to change is likely to be tough. In the largest developing country—the People's Republic of China—a similar process is taking place, demonstrating that the new advances present immediate opportunities not only for already industrialized countries but for all nations.

In considering the triad, it is important to note that each of its three cornerstones faces problems. The United States retains its lead in the creation and development of the more important emergent technologies, and signs are that it will continue to do so for some time. But the size of the federal budget deficit and the size of the trade deficit, as well as the process of deindustrialization in many traditional sectors that were once the powerhouse of the U.S. economy, are surely causes for concern.

Japan is exceptionally good at exploiting the new technologies and creating large-scale applications for diverse markets. Yet the Japanese, too, are seriously worried, as can be deduced from Japanese reports calling for improved economic and scientific strategies. There are several reasons for their apprehension. Their economic success has been built on an excessive dependence on exports. Profits have been reinvested in industry at home, and the resulting overcapacity has spurred in a vicious circle the need for an even better performance abroad. Given the Japanese people's high propensity to save, the domestic economy is finding it increasingly difficult to consume the income they generate. Meanwhile, the Japanese government's inability to redress the country's chronic balance of payments surplus leads to recurrent threats of retaliation from exasperated, less competitive trading partners.

The yen/dollar exchange rate implies that Japan has the highest per capita income in the world, yet few would deny that the living standards of ordinary people do not reflect this fact. Part of the production capacity devoted to promotion of exports needs to be switched to expansion of social infrastructures and improvement in the quality of life. The housing stock, the environment, and infrastructures in the less favored regions are all in need of upgrading.

With an economy long oriented toward "creative copying" and finding applications for advances achieved elsewhere, Japan admits a lack of individual creativity among its people, especially in the basic sciences. This is a by-product of a culture and an education system that instill virtues of obedience and teamwork rather than initiative and individualism. The future of Japanese technology must be based on independent effort in fundamental research and not on the import of technology from more advanced countries, as during the century-long process of catching up that began with the Meiji Restoration. Savings and consumption patterns will have to alter. All this is likely to mean major changes in the education system, a new role for the young in what has been a traditionally hierarchical society, and wider opportunities for women (still a significantly smaller part of the labor force in Japan than in any other industrialized country).

Western Europe, on the other hand, appears less oriented toward the future. On the whole, the economies of Western European countries are less concentrated on advanced sectors and are more balanced in their strengths. High-tech sectors are not the most aggressive elements in their economies, even though some of these sectors constitute areas of strength—nuclear energy, aerospace, and robotics. Overall, Europe is too weak in certain critical areas of microelectronics and information technology—for example, in basic electronic components, very-large-scale integration technology, and supercomputers. The most negative aspects of the situation in Europe are a lack of cohesion in many emergent sectors, inadequate infrastructures, and a dispersed and fragmented market.

Europe's cultural heritage, its deep-rooted traditions in the arts and craftsmanship, and the availability of welfare provisions—care and assistance for the individual citizen, typical of the "welfare state"—are equally distinctive characteristics. They give European nations an edge over the United States and Japan in applying new technologies to traditional industrial and services sectors and in creating diversified, personalized products in response to market needs. Productivity of labor has risen in Europe, although to the detriment of full employment, and so has product and process flexibility. Europe's reputation for quality products is being maintained increasingly through the adoption and adaptation of new technologies in their production.

Globalization is moving faster than the long-heralded political and economic unification of Europe. Global competition came about suddenly, and it caught Europe off guard. These two unifying processes—on the one hand, the European Economic Community (EEC) and, on the other, the global economy—are now developing side by side; in some areas they are competing. Where the European firm is an acknowledged leader in an advanced sector, these processes run in tandem; where the reverse is true, European considerations tend to take second place.

Many European firms are seriously at risk of being left behind in this competition by becoming the weak link in the triad, a link that provides ideas, labor, services, and markets but essentially leaves strategic initiatives to their U.S. and Japanese partners. Europe is a divided continent and, considering only the EEC, an uneasy mix of old, established, industrialized countries and others in which rural cultures and outlooks still prevail. Policies to pump subsidies into ailing agriculture, declining industrial sectors, and overstretched nonmarket services such as public sector health care, road and rail networks, postal services, and primary and secondary education—Europe's first response to the economic crises of the 1970s—are proving difficult to remove.

Basic scientific research is still in good shape in Europe, and individual scientists and relatively small, high-level research groups produce excellent results. The few large, cohesive research teams that were created in Europe in certain areas of scientific research, such as the European Organization for Nuclear Research (CERN) in high-energy physics, are highly competitive. Europe even occupies a leading position in some important industrial sectors: precision machine tools, electronic instrumentation, pharmaceuticals, and fine chemicals. In general, however, European industry still tends to think in terms of closed markets with the survival, wherever possible, of producer cartels. Public procurement policies remain largely at the level of single nations; this is a serious obstacle to a more active, relevant role in the world economy. There are, however, heartening signs that Europe is becoming more aware of its weaknesses in this area. Initiatives in science and technology are being undertaken at the EEC level and, separately, in the ambit of the so-called EUREKA program of coordinated, transnational research and development in advanced sectors.

An interdependent and more open world society will lend itself best to the challenge of innovation. The world needs much more material growth; the world population has reached 5 billion and will increase to 8 billion in 2050 before it stabilizes at something under 10 billion. The increase will take place almost entirely in the Third World. A quarter of the world's population now inhabits today's industrialized countries, but this proportion will fall to less than 20 percent in 50 years. The inhabitants of industrialized countries already consume three-quarters of the world's energy and mineral resources. It is difficult to imagine that disparity on this scale can continue far into the next century.

It is essential for world society that the existing gap between North and South be narrowed. This narrowing should be seen not only as a moral obligation for prosperous nations but also as in their own long-term interest. Development in the Third World will create areas of complementary production that will expand and broaden the international economy. This will, in turn, generate new markets for tradable goods and services, thus replacing today's frenetic paper market in financial instruments. If present trends continue, this market is bound to increase the disparity between the rich and the poor in the world and hamper investment in industry and other productive activities.

Patterns of development for the Third World need not follow those set by today's industrial economies. Available new technologies (for example, in agriculture, rural industrialization, and education and for the delivery of services) make it possible to achieve a more balanced growth without the exaggerated and disorderly urbanization and subsequent unemployment and other social ills now occurring in much of the Third World.

In this optimistic vision of the future, multinational enterprises are very important, but not in the traditional sense. Globalization will be increasingly linked to innovation. Furthermore, many small and medium-sized multinational corporations will emerge, relying on alliances that draw on the experience and information available to partners in each market in which the alliances operate. The role of government will not diminish. This role will not necessarily be antagonistic but will provide overall strategic direction, infrastructure, monitoring of conditions for fair competition, and preservation of cultural heritage and environmental quality.

Thus, the availability of abundant raw materials and cheap labor are no longer key factors for success in the world market. New technologies restore vitality to certain sectors in industrialized countries, sectors that were hitherto viewed as almost certain candidates for relocation to the Third World. At the same time, developing countries now have available to them a whole set of new technologies that lend themselves to blending with traditional technologies and thereby make faster development possible across the board.

Those developing countries endowed with raw materials and energy may convert them into more valuable commodities, but unless they are able to master the technology needed to upgrade such commodities, they will derive little benefit from this primary transformation. Emphasis must therefore be placed on research and development and enhanced international cooperation, because it is not in the interest of advanced countries to keep the developing countries' margins so low as to hamper their advancement and preclude their becoming healthy producers and active market forces. Whether this happens depends largely on the wealthier societies of North America, Western Europe, and Japan. Responsibility therefore lies with them.

REFERENCE

Ohmae, K. 1985. Triad Power: The Coming Shape of Global Competition. New York: Free Press.

Global Flows and Barriers

Trends in International Technological Cooperation

GERALD P. DINNEEN

GLOBAL TECHNOLOGY has had a tremendous impact on the structure of industry. As technological advances occur ever more rapidly and in many parts of the world, industry has responded by forming joint ventures, alliances, and overseas research and development (R&D) organizations. In turn, these new organizations expedite the globalization of technological innovation. This interaction will continue.

Moreover, industries have had to work in the global marketplace to get a proper return on investment for ever more expensive technological developments. They have formed international marketing organizations. In addition, industries have gone to worldwide production, both to reap the benefits of lower production costs and to lessen trade tensions in countries where imports have had a significant impact on domestic industries. There has also been transfer of technology through licensing, joint ventures, and the creation of new business entities. These have increased the number of subsidiaries of the multinational corporations and have created new strategic alliances.

This has been made possible by a number of communications technologies, such as fiber optics, electronic mail, and particularly satellite communications, a truly revolutionary innovation providing communications worldwide, including developing countries.

This trend toward the formation of joint ventures and international teams for major projects will continue. In fact, one change for the global economy is increased cooperation earlier in the development cycle in industrial research

These papers are adapted from the transcripts of the Sixth Convocation of the Council of Academies of Engineering and Technological Sciences.

and development. Cooperation is already prevalent in academia, but not in industry, except for joint ventures.

In the past, industry has not favored cooperative research and development because of concern about antitrust violations and loss of proprietary advantage. Obviously, neither of these considerations has inhibited cooperative research and development at universities.

Since the easing of antitrust regulations for cooperative precommercial research and development and the pressures on industry due to increased worldwide competition, there are indications of a movement toward joint technology efforts. In the United States, there is a Microelectronics and Computer Technology Corporation (MCC) made up of about 20 companies that are cooperating in the precommercial development of technology. Examples in Europe include the European Strategic Program for Research and Development in Information Technology (ESPRIT), the Alvey Directorate and associated activities in Great Britain, and certain projects of the European Economic Community (EEC). In Japan, there is the fifth-generation computer project and other activities sponsored by the Ministry of International Trade and Industry (MITI). Although it is still early, certain examples show promise that these consortia and collaborations will yield good research and development results.

The true test of the success of these joint activities is whether or not research and development will find application in the companies that are supporting it. Will it be possible to transfer technology from these joint activities back to the companies? Although there are encouraging signs, that answer is still an unknown.

One major impetus for cooperative activity is that it is becoming increasingly difficult for companies to hold on to their competitive edge. This difficulty is due to the rapid diffusion of technology; the obsolescence of existing facilities and the high capital costs of new facilities, particularly in microelectronics and some of the new fields; the complexity of scientific and technological endeavors that are too big for any one company; economies of scale; and the discontinuity between R&D and commercialization of research results.

These problems could be mitigated through increased collaboration, but there are barriers to cooperative efforts. The most serious of these is the cultural barrier, or the difficulty that engineers and scientists have in accepting technology from an external source, let alone another nation. This "not-invented-here" syndrome is a strong phenomenon in the United States. It appears to be less strong in Japan, for Japan has been extremely successful in transferring U.S. technology and applying it to new products.

Protectionism is a barrier not only when it is instituted by government but also when it exists between companies. The latter is due to a company's desire to retain proprietary advantage. However, with the rapid diffusion of technology, it is becoming more and more difficult to maintain proprietary advantage.

The desire to preserve domestic employment and to maintain domestic production for reasons of national security are important barriers to international technological cooperation. Another large barrier is the need for standards, open communication, and better protocols for transmission of data between computers and other electronic devices.

Even within any one country such as the United States, where large computer companies and small semiconductor companies may have different objectives and backgrounds, the gap between the various participants in a technical field constitutes a barrier to cooperation. The problem is aggravated one hundredfold when there is collaboration between nations. Differences in economic strength, culture, and work force must all be overcome.

One way to overcome these barriers is to establish truly international laboratories. One such facility is IBM's laboratory in Zurich. Although Honeywell's corporate research laboratories, which employ 450 people, are not international in the same sense, 42 percent of the new scientists and engineers hired in the last 2 years are foreign nationals. This high proportion of new hires who are foreign nationals is not unusual when one considers that 55 percent of the U.S. graduate students in engineering are foreign nationals. This is the beginning of another kind of international cooperation.

Whichever form international cooperation takes, it will provide for a free flow of technology, which will in turn create more technological innovation to fuel the global economy.

Technological Cooperation in Europe

HARRY L. BECKERS

WHEN INDUSTRIALISTS AND ACADEMICS get together to talk about science and technology, they often need to spend some time sorting out their respective definitions of those terms. There are many differences between industrial R&D and university R&D. The objective of university R&D is to deliver the very good engineers needed in industry and government. Evaluation of the quality of university R&D is frequently accomplished through the peer review of articles that result from these endeavors and that are published in the open literature. Support for university research is determined by a worldwide equilibrium between the availability of government funding and the needs of academic research establishments. When research establishments in a particular country think that they are not getting enough money from the government, they exert pressure in appropriate places by comparing themselves with other countries. If the politicians think the academic institutions are getting too much, they always manage to cut back the budgets for university R&D.

R&D in industry, on the other hand, is judged by the bottom line. If the research results are translated into products that make a profit for the company, that research effort is considered worthwhile. In general, research in industry is done for competitive reasons.

Rather than publishing the results in the open literature, industry uses patents and licenses as a system for the transfer of technology worldwide. Technology can be sold, or it can be used to the originator company's advantage. In other words, the system in industry is different from that in academia. At learned societies one of the questions is always how many articles, rather than how many patents, have you produced? It is important to make this distinction between academic and industrial research when talking about cooperation, technology transfer, and government subsidies.

One impetus for increased technological cooperation in Europe is competitive pressure from the United States and Japan. In the United States a great deal of money is spent on industrial research. In addition, defense spending subsidizes much U.S. R&D. At one time, 60 percent of the R&D budget at the General Electric Company was paid for through defense contracts. Although such support for R&D is all done on a commercial and contractual basis, this definitely gives a company an advantage in the resources it has available.

Japan is moving toward having more basic research conducted by universities. In fact, four or five universities have arranged to conduct research in the laboratories of Toshiba, Matsushita, and other large companies. This is a clever way of introducing basic research in a country and making sure that the results are used by industry.

In response to these activities in the United States and Japan, the EEC Commission in Brussels is attempting to organize Europe for collective efforts in R&D. Currently, a "common market" does not exist, nor do people from one country always know what the people in other countries are doing. Some EEC programs are attempting to reduce this technological isolation by bringing together researchers from different countries, for example, from a French university and a company in the Federal Republic of Germany, or a British company and a laboratory in the Netherlands. The amount of money involved is relatively small, but it has a tremendous impact on the relationships within Europe.

A committee to provide advice to the EEC programs is the Industrial Research and Development Advisory Committee (IRDAC), consisting of members from European industries with widely differing backgrounds. IRDAC was set up as an independent committee in 1985 by then EEC Commissioner Vicomte D'Avignon, in an attempt to ensure that European R&D programs would have an adequate industrial orientation. Through working parties, IRDAC has involved a wide spectrum of industrial experts, thus ensuring that a balanced view of European industry forms the basis of IRDAC's advice to the Commission. IRDAC replaced an earlier advisory committee made up of member-state representatives, because it was felt that the advisory role of the committee was hampered by political considerations.

Five or 10 years ago the governments were spending considerable sums of money on the sunset industries, and now, they are putting much of their money into high-tech, or sunrise, industries. In the EEC R&D program, there is also a project called BRITE, Basic Research in Industrial Technologies in Europe, which is aimed primarily at the sunshine industries, which are perhaps less high-tech but are still earning money.

Through this endeavor, industry from one country works with a university in another country. IRDAC closely monitors this program and has established several working parties consisting of industrial experts to provide advice on relevant activities in, for example, the materials and mechatronics fields.

Although EEC funding levels may be lowered, funding of national R&D programs in some EEC member countries has increased. This has created a paradox in Europe as individual countries strive to increase funding for their national R&D programs at a time when they are trying to reduce the funding for EEC programs, which frequently address the same areas of technology.

Europe's nationalistic tendencies have resulted in inefficient use of R&D funds. Countries claim to have a unique program designed to increase the competitiveness of a particular industry, while a neighboring country has exactly the same program for the same industry. A second problem is that small European companies need to be "Europeanized."

Europe still has a long way to go, but it may help put the situation in perspective to remember that just over 40 years ago we were still at war with each other.

Driving Technologies

Advances in Software

GEORGE E. PAKE

My responsibilities for the past two decades have been as a research manager in association with a research enterprise that has devoted substantial effort toward advancing software technologies. In our research and technology planning, it has often been difficult to partition the research—or especially the projected technological and business systems—into a software sector and a hardware sector. Indeed, some of our greatest technological successes were a consequence of taking the largest overall systems viewpoint: the recognition that the real systems issues were those of the supersystem, which integrates software and hardware so intimately as to obscure hypothetical boundaries between the software system and the hardware system.

As an example of these integrated software-hardware successes, consider that the human interface to the computer that was pioneered by the Xerox Palo Alto Research Center is now being copied or imitated by many computer workstations on the market. This human interface is characterized by a bit-map display, the mouse for guiding the display cursor, and "windows"[1] on the display screen. As is readily seen, the software for this interface requires certain associated hardware components (e.g., the mouse) and characteristics of the processor hardware to operate and refresh the display. Thus, the experimental Xerox Alto computer of the 1970s, introducing the bit-map display and WYSIWYG,[2] employed a larger portion of the computer memory and processing power to operate and refresh the display than had been used in most preceding computers. This architectural feature of the hardware was carried over into the Xerox 8010 Star system in the first commercial product offering to take advantage of the software for the Xerox human interface. This approach is now becoming almost ubiquitous for computer workstations, notably having been appropriated by Apple for the Macintosh personal computer.

A similar difficulty applies in searching for the boundaries of a software industrial sector. Portions of this possible sector are readily recognized. The publishers and developers of software for the personal computers that are finding wide application in the business world are surely part of this industrial sector (e.g., Microsoft or Ashton-Tate). So also are the firms that deal in applications software for large computer systems (e.g., Electronic Data Systems). A substantial fraction of the business of the major computer companies (IBM, DEC, Fujitsu, etc.) is in software; yet these companies are frequently regarded, and probably regard themselves, as hardware companies.

Despite the difficulties we may have in defining the software industrial sector, we can agree that software and associated software technologies are playing a larger and larger role in other industries. Thus, there is widespread interest in software technology and in its associated advances as they affect and stimulate the global economy.

Although software technology is widely held to be critically important to the broad technological advance that now spurs the global economy, it is difficult to define. The new Software Engineering Institute being established with U.S. Department of Defense funding at Carnegie Mellon University is, according to its First Report in *Carnegie Mellon Magazine*, struggling to define what software engineering as a professional field really should be. "Unlike other engineering professions, software engineering does not have well-articulated goals, standards, or methods for its practice," states the report.[3] I would add that the key scientific and engineering underpinnings of software technology remain to be identified and organized into a course of instruction so that we might proceed to train a software technologist in the same way we train a metallurgical technologist, for example. Many of these difficulties relate to the earlier problems we encountered in attempting to segment software from hardware. Furthermore, I believe that complete separation of software considerations from hardware considerations can be done only at severe peril to the effective function and efficient performance of the overall supersystem, hardware plus software.

Consistent with this view, my list of the key technological advances contributing to effective software technology begins with the components and architecture of the hardware systems used for software development. Of paramount importance has been the decline in cost and the increase in speed of processing power and memory, putting at the disposal of the software developer a fast, fully interactive system with rapid access to memory and to data bases.

These hardware advances have made possible the second key advance, the advent of extremely powerful software development environments that combine advances in writing, editing, running, and debugging of software. Gains in effectiveness are afforded by windowing, bit-map displays characterized as WYSIWYG, and other programming innovations that are them-

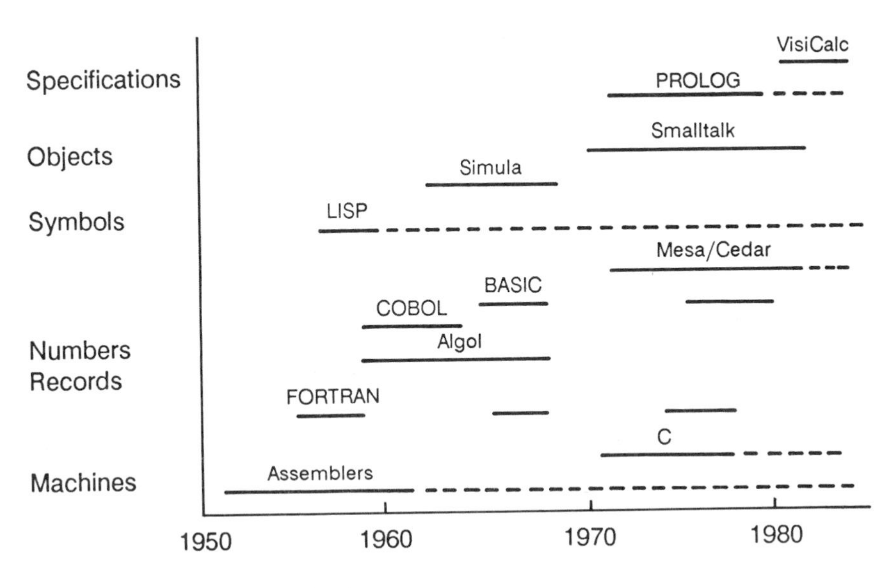

FIGURE 1 Selected programming languages.

selves inherently software in character but are enabled by the more powerful hardware systems.

A third area of technological advance is in the development of different kinds of programming languages, each designed to be optimal for certain types of problems. Many experts classify programming languages into three general categories: (1) procedural, such as FORTRAN, BASIC, COBOL, C, and LISP; (2) object-based, such as Simula and Smalltalk; and (3) constraint-based, such as VisiCalc and Prolog. See Figure 1 for some selected programming languages arrayed in these categories against a time line.

The fourth area of technological advance is, for software development projects, the systematic forward planning and task analysis—application of sound engineering principles to the system specification and to the handling of change orders (last-minute revisions) for large software development projects.

How can we control the powerful systems we establish, so that they do not enslave us or stifle creativity? I believe that, insofar as software development and the supporting technologies are concerned, we have not yet installed systems of such rigidity as to be threatening in that way. The software development process is extremely diverse throughout industry, and many of our concerns about its cost and low efficiency relate to that diversity. There are many languages and development environments in use—a plurality that fosters creativity and facilitates revolutionary or innovative approaches, but at the cost of low efficiency and low schedule predictability. The desire for greater efficiencies in large software development tasks will produce pressures for greater standardization in the languages used and in software engineering procedures. Once these pressures begin to rigidify software development processes, at some time far in the future, concerns about enslavement and stifling of creativity could indeed characterize software technology. My personal assessment is that software technology is today remote from that degree of standardization and rigidification of the development system.

ACKNOWLEDGMENT

I would like to express appreciation for helpful discussions with three of my Xerox colleagues, Dr. Adele Goldberg, Dr. Robert Ritchie, and Dr. Robert Spinrad. I also acknowledge a helpful and stimulating discussion with Mr. William Gates, chairman of Microsoft Corporation.

NOTES

1. Windows on the display screen are in many respects a metaphor for the way we use several papers and books on the desk as we work. Placing on the desk a book or paper to which we refer as we write a manuscript is, in effect, opening a new window to our visual attention. Of course, it may temporarily cover other books or papers to which we have been referring. In the context of the computer interface, windowing is setting aside an area of the display screen as a domain of visual attention, within which activities are carried on either alongside or over other windows on the display. The secret of opening these windows, storing other windows elsewhere on the screen, and carrying on independent computing or editing activities within each window lies, of course, with the programming. This approach is a major software advance of the past decade.
2. WYSIWYG is a nearly pronounceable acronym for "What You See Is What You Get" (meaning what you get on the printout from an associated laser printer). Each small spatial element of the printed page is accessed by the laser beam from the scanner as the beam is controlled by the bit-stream. Considerable design effort is made to see that the display, which is inherently of lower resolution than the laser beam-addressed page, closely resembles the page that will be printed on the paper. This page layout is typically laser-scanned at about 300 elements per inch, whereas the typical computer display has a resolution of 70–75 elements per inch. Considerable software effort is required to make the display screen appear to the knowledge worker just as the page printout will look from the higher resolution laser printer.
3. Maguire, M. 1986. First report from "The Software Capital of the World." Carnegie Mellon Magazine (4, Summer): 17.

Advances in Materials Science

PIERRE R. AIGRAIN

MATERIALS ARE SO IMPORTANT for society that we classify the various eras in the development of mankind by the kind of materials used; for example, we commonly refer to the Stone, Copper, Bronze, Iron, and Steel Ages. Only 120 years ago we entered the Steel Age; all the materials we now call new, such as aluminum, have come into being since that time.

New materials are discovered and introduced practically every day. In 1900, aluminum was still a specialty product. Plastics were not unknown; celluloid and bakelite were discovered at about that time, but their use was extremely limited. Celluloid was initially introduced as a substitute for ivory for billiard balls, and this was its main use for 20 years. Today, a comprehensive list of materials would include at least 500 entries, ranging from structural materials to conductors.

Although new materials appear every day, a long period of time elapses after their initial introduction and specialized application, which require only small quantities of the materials, before they become economically important materials in tonnage and value. Obviously, when materials are introduced for large-scale application, their influence on industry and society is enormous.

The sequence of events from introduction to large-scale use of new materials raises three questions that can be addressed first in a general, almost philosophical way and then examined further by looking specifically at the new superconducting materials. The first question is, Why have new materials been discovered so quickly during the past few decades, and will this trend continue? Will we continue to invent new types of materials at the same rate? The second question is, Why does it take a long time for materials to be introduced in all phases of industry? Why are they often limited to special applications for extended periods? The third question, of course, is, How can new materials really change industry and society when they finally reach the broad applications stage?

The answer to the first question is that new materials will continue their rapid trajectory of discovery and limited application for the following reasons:

This paper is adapted from the transcripts of the Sixth Convocation of the Council of Academies of Engineering and Technological Sciences.

A much better interaction between basic science and materials technology has developed. One important contribution of basic sciences to materials development has been the development of analytic instrumentation, including synchrotron light, for the study of materials. For centuries, the approach to materials research was systematic empiricism because there was no other way. Then there was movement to educated empiricism in which some general rules drawn from the basic sciences were useful.

The market has also been a strong influence on the rate at which new materials have been discovered. Sometimes the market demands a new product, which technologists must create. At other times, the technology is available, but there is no market for its applications. Optical fibers are a typical illustration of this interaction between technology push and market pull. There was a market need for optical fibers because of the crowding of the broadcast bands for radio communication and the lack of space underground in large cities to lay telephone wires. Because of improved scientific understanding of the light absorption process in solid materials, people realized that the absorption coefficient of light in a glassy material was almost entirely due to extrinsic problems, such as impurities. These could, at least in theory, be eliminated. Basic science research had also shown that intrinsic absorption of this silica was many orders of magnitude lower than that of the glass that was available only 15 or 20 years ago.

The technology with which to purify silica was available, and it could be transferred with few changes from the semiconductor industry, which had already developed chemical vapor deposition techniques. In addition, many ancillary components were available for building an optical fiber communication system, components such as the semiconductor laser, detectors, and signal treatments. All of these conditions combined to produce unexpectedly rapid development of new materials used for fiber optical communications. In fact, in this case, optical fiber has become an industrial product in an astonishingly short time.

Whereas materials technologies once developed independently, they now develop through continuous interaction between basic science and materials research, and this trend will continue. The interaction among technologies has become the biggest engine of technological development during the past few years.

The second question concerns the length of time required for new materials to achieve broad applications. For example, carbon fibers are used in the aeronautics industry and in the leisure industry for tennis rackets and fishing rods. However, it will take a long time before carbon fibers take the place of many automobile components.

There are several reasons for this phenomenon. Most of the new materials of interest are not costly, because they are based on plentiful elements of the periodic table. For example, new magnetic materials are based on iron and boron, which are very inexpensive. Neodymium is one of the second most abundant elements, and there are enormous deposits of bastnaesite and monazite from which neodymium is easily extracted. Despite the low cost of the elements themselves, at the present time the process for making these materials in the proper form is very costly. It often takes time to develop methods to reduce the price, and as long as a process is costly, the material will not be used in large quantities. In addition, the manufacturer of a particular material sells his product to people who make components, which are introduced into products that are then sold to consumers or industry. Consequently, there is a small profit incentive for the developer of the technology, because he is under the price pressure of a long train of customers.

Superconductors illustrate many of the problems inherent in bringing a new material from discovery to widespread application in a short time. Although superconductors were discovered before the First World War, they were not understood. Nevertheless, it was recognized that a material that could carry electricity without loss of energy should be extremely marketable. When it was discovered that these properties held true only for very low current levels and at very low temperatures, enthusiasm died out.

Solving the problem of the temperature and current constraints on superconductivity became a matter for pure science research. Some progress was achieved through empirical theories such as London's and later through the work of Bardeen, Cooper, and Schrieffer. New materials, such as niobium nitrate, that had slightly higher transition temperatures were discovered. Then Matthias applied his genius to devising more or less empirical rules, which led to the discovery of materials with higher transition temperatures, culminating at 23.2 K. Although this was an important discovery, it was not sufficient to maintain market pull.

Other people were working in completely different materials and found that some oxides, for example, have transition temperatures above 10 K. During that time there was also work in France on the lanthanum copper oxide-based material and yttrium copper oxide. These materials produced higher conductivity but were not tried at low temperatures.

Müller and Bednorz, working at the IBM Laboratory in Zurich, proposed that these oxides might work at a higher temperature. In early 1986, Müller and Bednorz published results showing the initial appearance of superconductivity at 40 K, a sudden jump by a factor of 2. Within three months, it was announced in the People's Republic of China that beginning superconductivity at 73 K and zero resistance at 80 K had been observed. The result was soon reproduced in the United States, Europe, and Japan. There are indications that a sample was superconducting at a temperature of 244 K, very close to room temperature. Thus, from the time of the first theoretical treatment of superconductivity in the 1930s, it has taken more than 40 years for superconductors to be brought to the stage where it is realistic to talk about widespread application of this technology.

Let us consider likely potential uses and effects of these new superconducting materials in answering our third question: How can new materials change industry and society when they reach the broad applications stage? First, they will be used in specialized, high-value-added applications in the electronics industry. The number of possible applications is enormous, from measuring small magnetic fields to fast computing with Josephson devices. These are applications that will come quickly.

Second, high-temperature superconductors will probably influence electrotechnology, that is, the production and transport of electrical power, even though several critical developments are necessary. For example, because the materials are ceramics, which are notoriously nonductile, it is difficult to use them as windings in electric generators. How can they be put in the right shape? What kind of auxiliary equipment is necessary to keep them at a low temperature? Even if they are superconductive at room temperature, their current-carrying capacity will be only one-fourth of the transition temperature.

Thus, ''high-temperature'' superconducting materials are much easier to use than those cooled with liquid helium, mostly because they can be cooled with liquid nitrogen, which is a much better coolant with a much higher heat of evaporation. However, enormous developments are required before these superconductors can be used for massive applications.

Information Technologies in Industry and Society

LARS RAMQVIST

BENJAMIN DISRAELI ONCE SAID, "The most successful man is the one who has the best information." This remark summarizes the business of information technologies—the production, processing, storing, communication, and use of information.

Information technologies have resulted in the development of one of the world's largest industries. Global production of electronics equipment in 1985 exceeded $400 billion, as consumption of semiconductors neared $25 billion (see Tables 1 and 2). By 1990 these production and consumption figures are expected to expand to at least $600 billion and $65 billion, respectively.

Today, cutting-edge technologies such as computers, software and artificial intelligence, fiber optics, networks, and standards have an immense impact on information technologies. Among the many applications of information technologies, three of particular importance are traditional telephony, mobile cellular telephony, and data processing and communication. Information technologies, in turn, affect many industries and society as a whole.

THE INFLUENCE OF CUTTING-EDGE TECHNOLOGIES

Very-Large-Scale-Integration Technology

Recent achievements in information technologies build on a rich history (see Table 3). The development of chip technology, for example, has been exceptional over the last three decades. Today, a million or more transistors can be included in one chip. In fact, the number of devices per chip has increased by 100 times per decade since 1958. If this pace of development could be applied to the automobile industry, it has been estimated that six

TABLE 1 Production of Electronic Equipment in 1985 (billions of U.S. dollars)

Industry	United States	Western Europe	Japan	Rest of World	Total
Data processing	80.4	21.2	17.3	9.9	128.8
Communications	28.6	17.7	7.5	4.6	58.4
Industrial	34.9	17.8	9.5	4.3	66.5
Consumer	16.2	10.1	36.1	12.4	74.8
Military	49.2	11.0	—	1.0	61.2
Transportation	8.5	2.0	2.8	3.0	16.3
Total	217.8	79.8	73.2	35.2	406.0

Rolls Royces could be put on the head of a pin and that each of them would cost about $3.00, give over 3,000 miles to a gallon of gas, and have enough power to drive the *Queen Elizabeth II*.

In very-large-scale-integration (VLSI) technology at the cutting edge of development, there are challenges in feature size, design complexity, and facilities for production. VLSI technology today includes feature sizes of less than 1 μm on the chip. On a biological scale, this is in the range of red blood cells and yeast cells to the smallest bacteria. However, feature dimensions as small as the human immunodeficiency (HIV or AIDS) virus, which is about 1,000 Å, have still not been reached (see Figure 1). The smaller the feature size, the faster the processing capacity and design complexity of the chip. Thus, the feature size is critical for the price–performance development in microelectronics.

The equivalent of hundreds of worker-years is now put into the design of a complex chip roughly 40–100 mm^2 in size. This implies making full use of advanced computer-aided design technology, including cell libraries and macrocells—that is, tools by which defined and tested blocks, such as a

TABLE 2 Consumption of Semiconductor Components in 1985 (billions of U.S. dollars)

Industry	United States	Western Europe	Japan	Rest of World	Total
Data processing	3.7	0.9	2.6	0.4	7.6
Communications	1.4	1.2	0.8	0.2	3.6
Industrial	1.5	1.1	0.8	0.2	3.6
Consumer	0.7	1.0	4.0	0.8	6.5
Military	1.5	0.4	—	0.0	1.9
Transportation	0.8	0.3	0.3	0.1	1.5
Total	9.6	4.9	8.5	1.7	24.7

TABLE 3 Key Achievements in Information Technologies (with inventors noted as appropriate)

Date	Invention	Inventor
1455	Gutenberg press	Johannes Gutenberg
1844	Telegraph	Samuel F. B. Morse
1876	Telephone	Alexander Graham Bell
1889	Strowger selector	
1901	Transatlantic wireless telegraphy	Guglielmo Marconi
1906	Triode vacuum tube	Lee De Forest
1910	Teletraffic theory	Agner K. Erlang
1923	Telephotography, the iconoscope	Vladimir K. Zworykin
1927	Feedback amplifier	Harold S. Black
1930	Telex, the coaxial cable	
1937	Pulse code modulation	Alec H. Reeves
1937	Xerography	Chester F. Carlson
1946	ENIAC computer	John W. Mauchly and J. Prosper Eckert
1947	Transistor	John Bardeen, Walter H. Brattain, and William Shockley
1958	Integrated circuit	
1962	Telstar satellite	
1965	Stored program control switch	
1966	Step index optical fiber conception	K. C. Kao
1970	Optical fiber lab test	Robert D. Maurer
1971	Microprocessor	
1976	Fiber optical transmission	
1990s	Possible software breakthrough with declarative languages	

processor unit or a memory function, can be used to build more complex designs. As chips become more complex, the trend is toward larger chips—even three-dimensional circuits—and optimization of the feature interconnect.

More limiting than feature size is the interconnect, and any drastic development in chip complexity will depend on a solution to this problem. Since the transistor or gate functions can be produced in small feature sizes, less than 1 μm, the interconnects between these features need to be of an equivalent size. However, high packaging densities and sizes that are too small will introduce such problems as increased resistance and decreased speed in the functions of the chip.

The frontier products of the late 1980s will be the 4- and 16-megabit (Mbit) dynamic random access memory (DRAM) chips, with the latter being only in the prototype stage over the next few years. In the next 8 to 10 years, however, DRAMs larger than 100 Mbits will be available on the market.

Producing these complex chips requires production facilities with a su-

Technologies (feature sizes in microns)	Examples for scale	Dimensions (microns)
	Human hair	10
	Red blood cell	7
1978 (5μ) CMOS/BIP →	→	
	Yeast cell	1
1988 (0.2μ) CMOS/BIP →	→	
1990s (0.2μ) Quantum Coupled devices (III - V) →	→	0.2
	AIDS virus	0.1
	Atom	0.0002

FIGURE 1 VLSI development (CMOS = complementary metal-oxide semiconductor; BIP = bipolar).

perclean atmosphere. In normal hospitals, it is possible to achieve a class 10,000 in the operating rooms, that is, 10,000 particles per cubic foot. Currently, class 100 conditions are normally applied in chip production. However, the most advanced production areas in the VLSI industry of today must provide class 10 conditions, the demand for which will be a driving force into fully automated production without people in the labs. Automated production will gradually be introduced during the next decade.

Nevertheless, VLSI technology still needs people; they are the key factor to success in this area. For example, in the United States, a relatively small number of skilled designers and engineers drive the $6 billion integrated circuit industry today.

Computers

Computers are one area where VLSI technology has been fully adopted. The development of digital computing started with the ENIAC project in the 1940s, followed by the machine productivity stage started by the IBM System/360. During this stage the number-crunching capacity of computers was steadily increased. In the 1980s we have only begun the third step of development—simplifying information handling and user interfaces.

This development has been made possible by advances in processing power, defined in millions of instructions per second (MIPS), in the central processing unit. A modern supercomputer, for example the Cray XMP, has a maximum capacity of 200–240 MIPS (see Figure 2).

VLSI technology has also dramatically increased the processing power in individual workstations. Three MIPS are already being used in commercial workstations and 100-MIPS workstations will be available in the early 1990s.

This processing power can now be used to solve problems at the human–machine interface. In the 1990s it seems likely that 80 to 90 percent of the computer's power will be used to support user interface functions.

New computer architectures driven by evolving software technologies will make more efficient use of the VLSI technology. Examples of such architectures are reduced-instruction-set computers (RISC) and various parallel processing arrangements, which might lead to a 10-fold increase in efficiency.

Software and Artificial Intelligence

The revolution in the software industry is still to come. We often think of the "software crisis" as a problem that can be solved only through the efforts of millions of programmers. But developments in hardware offer an opportunity to solve the software problem by combining good tools with engineering skills.

The overall development of software technology can be described as a series of discontinuities. The technology of computer languages has progressed from the machine level, where programs are written more or less as ones and zeros, to the assembler level, which affords the programmer a somewhat friendlier notation but still requires the programmer to describe the logic of the program in the same minute detail. Next are the so-called high-level languages, such as Fortran, COBOL, Pascal, and Ada, which imitate the languages of mathematics, accounting, or whatever the application area is and allow the programmer a higher level of expression. However, these languages still require the programmer to explain exactly how the computer is going to solve a problem, and therefore, the order of the lines

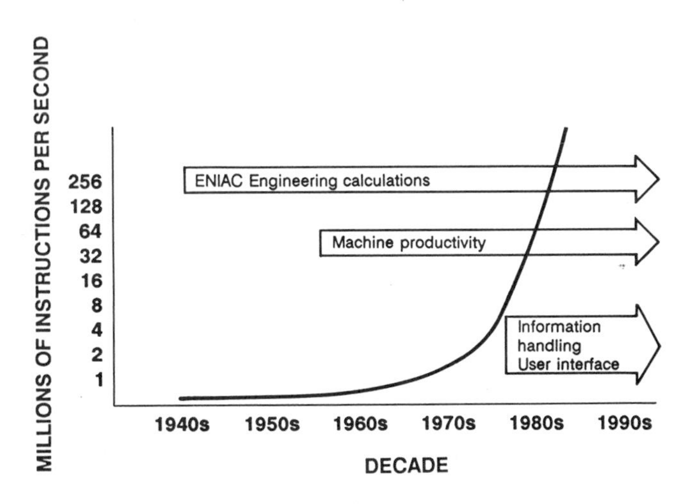

FIGURE 2 Digital computing development.

of code is very important. Finally, there are very high level or application-oriented languages, such as LISP, Prolog, and other so-called fourth-generation languages (4GL), in which the programmer declares what the computer should do, not how it should do it.

All the "how" languages, up to and including Ada, are called procedural, or imperative, languages. The very high level languages are declarative, or applicative, languages. They differ from the "how" languages in clarity, suitability for parallel execution, computing power requirements, and applications.

Declarative languages are usually more concise and clearer than procedural languages. They are also intrinsically suited to parallel execution, whereas procedural languages can exploit parallelism in a problem only with great difficulty. A drawback of declarative languages, until now, has been their need for substantial computing power, or preferably a new computer architecture. Procedural languages, on the other hand, are quite efficient in traditional "von Neumann" computers.

Partly because they need so many MIPS, declarative languages have found relatively few applications in industry until now. But there are reasons to believe that this is about to change—that the procedural languages are like the dinosaurs, growing larger and larger toward their extinction, and that the present declarative languages are like the first mammals, still small and hiding in the bushes, but poised ready to take over the world.

Language is only one factor that influences software efficiency and quality. The methods and tools used to support software development and handling are as important as the structure of the hardware and software.

In the telecommunications industry, large real-time systems with software written in millions of lines of code are needed to support our public switching system (AXE). These data bases sum up to more than 400 gigabytes. Another way to think of the size and complexity of this system is to consider that we have installed more than 10 million telephone lines in more than 50 countries.

The software content of a single AXE installation is on the order of 2–5 megabytes, and there are numerous versions to meet different market requirements. This calls for very good tools for releasing different versions and updates. As a result, it is absolutely vital to use results from information technologies research. With large systems, use of these results, in turn, requires extremely good software management and planning, as well as new ways of structuring systems. Reusable software and different kinds of software tools for different parts of a computerized system are needed. Today new technologies are continually being introduced; for example, artificial intelligence technology could be valuable for creating a good human-machine interface to a "conventional" computer system (see Figure 3).

Artificial intelligence combines such mechanistic concepts as repetition, precision, and data handling, and then uses this combination in the broader

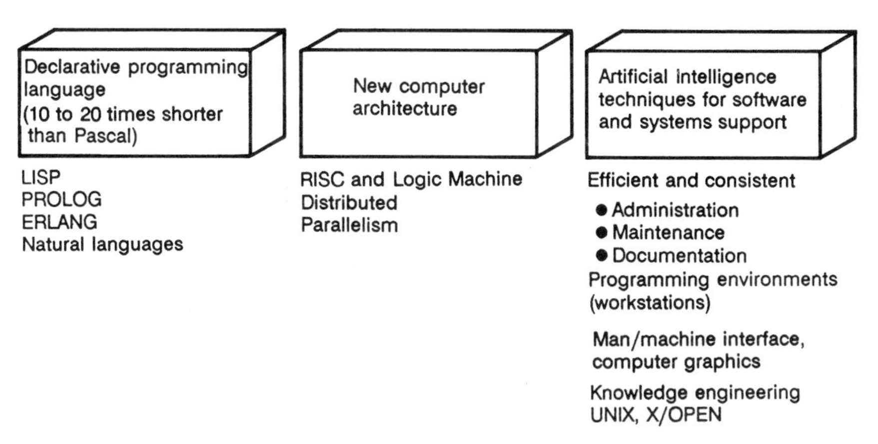

FIGURE 3 Software methodology and tools.

applications of expert systems and knowledge engineering. An expert system implies a combination of a knowledge base and data linked to a general problem solver. This is in sharp contrast to conventional programming where the data are processed by "hardwire application knowledge" programs, that is, where the algorithms are processed in specially designed hardware.

Prototyping and new languages and means for specification make the early phases of development more specific; this process is important because a major task in software development is the fundamental system design. But fourth-generation languages are not always the only solution. As cutting-edge technologies advance, it is becoming more and more important to begin developing standards, formal or de facto, such as those put forth by standards-setting organizations and in operating systems of large manufacturers. Adhering to standards allows an organization to concentrate resources in areas where it can add substantial value.

The next crisis in computing will be the need to handle the rapidly growing amount of information that will be available in distributed data bases. This poses many challenges for research. For example, we need new ways of describing data and classifying relationships between data and finding and retrieving data already stored in data bases.

Fiber Optics

Increasingly, more powerful computers and complex software and artificial intelligence need advanced communications. The solution is fiber-optic transmission. The key achievements in fiber optics and related industries began in 1970 with the development by Corning of optical fiber. During the 1970s complete fiber-optic telephone networks were already up and running in the

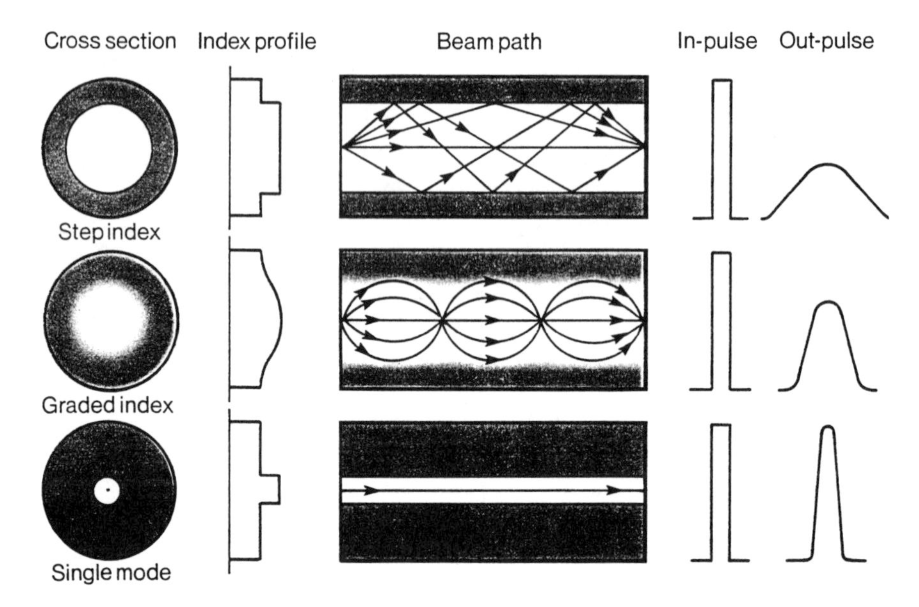

FIGURE 4 Types of optical fibers.

United States and Sweden. The commercial breakthrough came in 1984. By 1988 the Atlantic cable will be in operation. For the 1990s we expect a breakthrough in subscriber networks, the so-called local area networks (LAN), especially for high-definition television (HDTV) and data communications.

The technical development of optical fiber includes the step-index, the graded-index, and the single-mode fibers (see Figure 4). For advanced applications and broad bandwidth networks, the single-mode fiber will be used because of its remarkable capacity to transport narrow signals.

Dramatic achievements have been reached in the practical use of fiber-optic transmission systems. We will soon have the ability to carry a million or more telephone calls on one fiber over distances of more than 1,000 kilometers without repeaters (see Figure 5). By the end of this century, a fiber pair will probably handle 100–1,000 channels, each with a bandwidth of 1 gigabit per second (Gbit/s). This can be contrasted with coaxial cable transmission technology, which permits a maximum of 10,000 telephone calls per pair, with a maximum distance of 4 kilometers between repeaters.

To build a fully optical network, it is necessary to be able to switch light. One technology is the electro-optical directional coupler (EDC), with which light can be switched in frequencies of many gigabits per second. This coupler is made of lithium niobate ($LiNbO_3$), in which certain optical properties, such as the refractive index, change as a function of a magnetic field induced into the material.

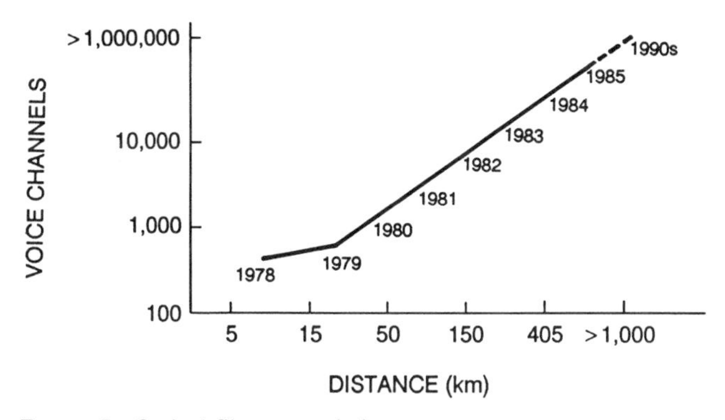

FIGURE 5 Optical fiber transmission.

Networks

With fast optical transmission tools, it is possible to build networks for the future. Let us look at the bit rates needed for telecommunication services. All the bit rates are in one way or another linked to the original 64-kilobit-per-second (Kbit/s) rate of normal telephone services. Although in the future we will see HDTV and complex inter-LAN/PBX communication systems, it may be possible to incorporate the transmission capacity of these systems in 140-Mbit/s systems. Many qualified image-compression techniques are needed to keep the bandwidth below 140 Mbit/s, but that is already almost possible (see Figure 6).

From today's telephony and data communication systems, broadcast TV, and terminal communications, we will see the gradual emergence of the integrated services digital network (ISDN), fiber-optical-based cable TV, and

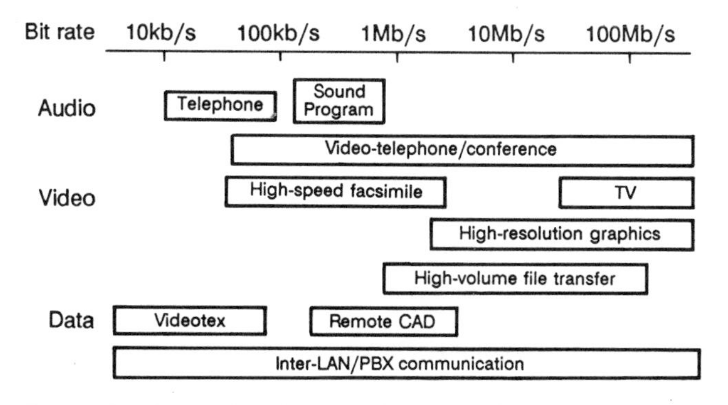

FIGURE 6 Bit rates for telecommunication services.

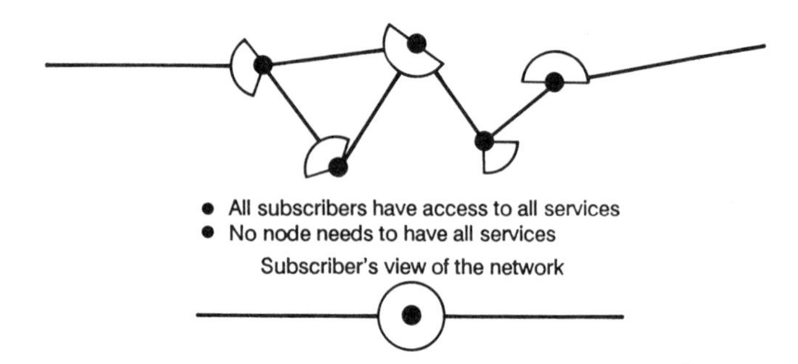

FIGURE 7 Functional distribution in future ISDN/broadband network.

LANs. Finally, all systems will develop into one broadband network, having a bandwidth of 140 Mbit/s as described above.

The future ISDN/broadband network will consist of intelligent nodes. The functional distribution in such a network will be such that the services for voice, data, text, and image could be divided between the different nodes in the network. The subscribers' view of the network, however, will be that they have continuous access to a fully equipped node and network for the services rendered (see Figure 7).

Standards

To be able to make full use of the information technologies discussed so far, one must communicate with other people, other companies, and other countries. Thus, standards are needed. The standards in information technology are based on open system interconnection (OSI). Figure 8 represents the outlook for future developments in information technology in seven layers, from the physical layer to the application layer.

APPLICATIONS OF INFORMATION TECHNOLOGIES

Powerful computers, advanced software, and fast and reliable communications represent all the basic technologies necessary for a modern information system. The ingredients of such a system will be the coexistence, connectivity, interworking, and standards of voice, data, text, and image with full reliability, availability, maintainability, and above all, ease of use. VLSI and computer technologies have made these conditions possible.

Three of the main applications of information technologies today are normal voice telephony, mobile telephony, and data communications. Regardless of ISDN and broadband network discussions and plans, much remains to be done in normal telephony before the world's needs will be met. For many years to come, a basic customer requirement will be voice telecommunications.

Normal Telephony

At present, normal telephony is the driving force in information technologies. The shift from analog to digital switching and transmission technologies will have a tremendous impact as the telephone network now becomes an integrated service network. It is also important to remember that the majority of the world's inhabitants do not have access to plain voice communication.

As shown in Figures 9 and 10, there were 454 million main lines in service worldwide in 1987 and 37 million lines in the local exchange market. These numbers appear large in themselves, but in relation to the number of potential users worldwide, they show another story (see Table 4).

Not surprisingly, a clear relationship exists between the telephone density and the gross national product per capita for the different countries of the world. Telecommunications are a vital part of the development of the infrastructure of a country. For many decades to come, there will be a genuine need for basic services in public telephony.

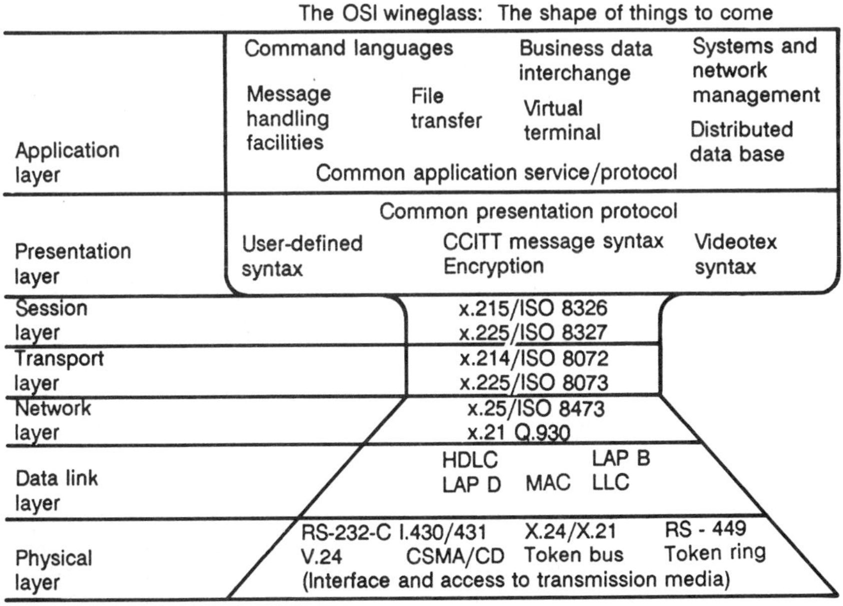

FIGURE 8 Standards in information technology (open system interconnection).

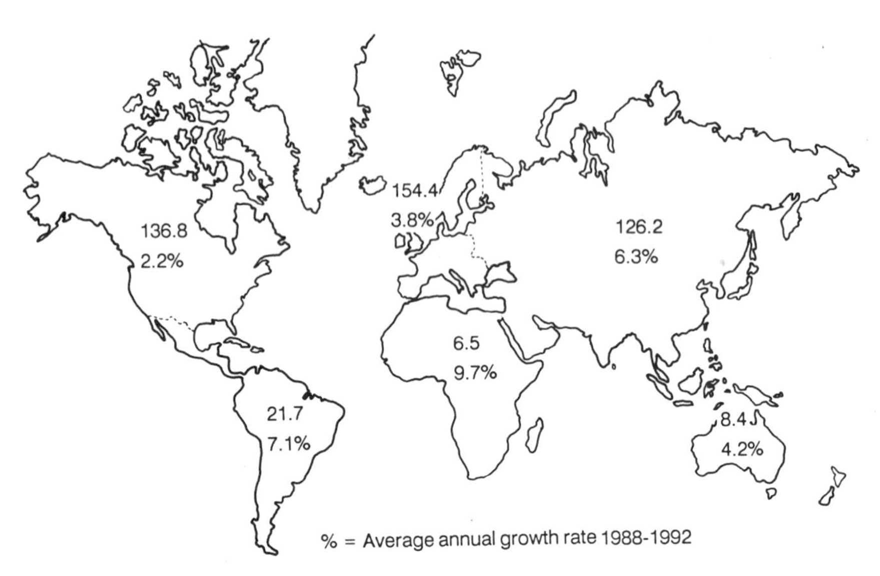

FIGURE 9 Number of main lines in services (millions). World total in 1987, 454 million (4.3 percent annual growth rate).

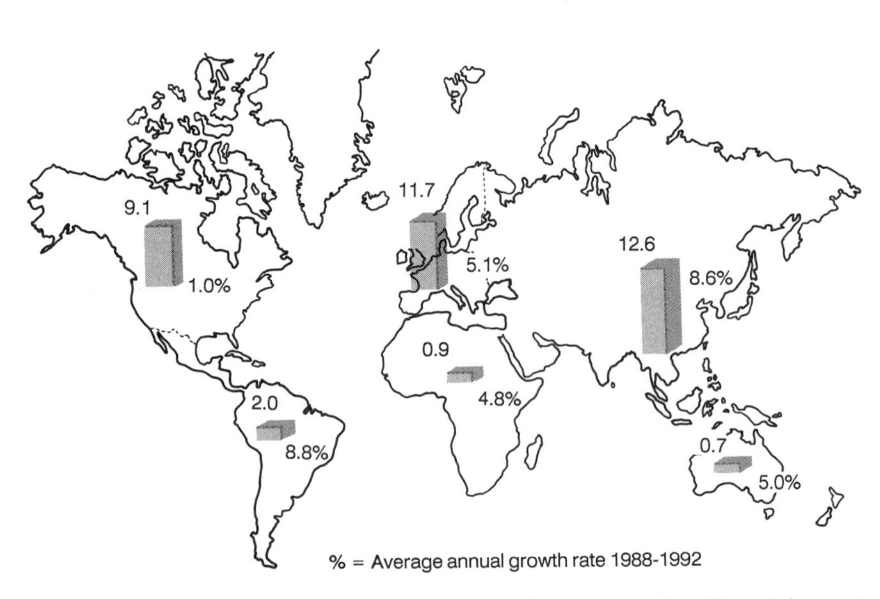

FIGURE 10 Local telephone exchange market. In 1987 there were 37 million (4.4 percent annual growth rate) installed local lines.

TABLE 4 Number of Telephone Main Lines in Service per 100 Inhabitants Worldwide

Country or Continent	Lines/100 People
United States	48
Australia	38
Europe	29
South America	5
Asia	4
Africa	1

Mobile Telephony

Basic voice communication has taken a big step forward with the mobile cellular telephone system, which allows customers to communicate by telephone while in transit. Mobile telephony techniques were not developed earlier because the infrastructure lacked the necessary advanced technology. Since the 1960s, manual open systems have developed into automatic systems, which in turn evolved into cellular systems with full flexibility and reliable continuous communications regardless of where the subscriber moves within the areas covered by the base radio stations. This development was totally dependent on advances in microprocessors, digital synthesis with semiconductors, and stored program control switch technology (see Figure 11).

In January 1987 less than 3 million mobile telephones were in service worldwide, with the highest use being in North America (see Figure 12). The cumulative growth forecast for cellular phones projects a total of 7.5 million by 1990, including 2.3 million in Europe and 3.8 million in the United States.

Mobile cellular telephony is still in its developmental phase, as shown by

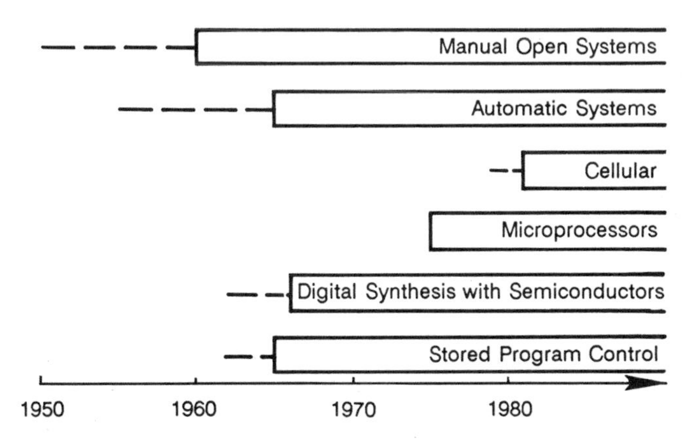

FIGURE 11 Background of cellular telephony.

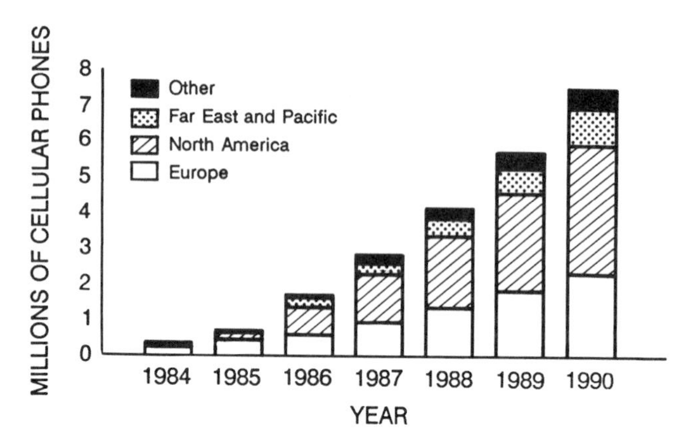

FIGURE 12 Cumulative growth forecast (cellular phones).

the numbers of phones per 100 inhabitants in 1986 (see Table 5). In urban areas the penetration of cellular phones is higher. In 1986, for example, the number of cellular phones per 100 people ranged from 2.5 in Stockholm, to 1.0 in London, to 0.1 in Tokyo. The degree of penetration is of course a function of time, as shown in Figure 13.

It is possible that in the early 2000s, digital mobile cellular telephony will play a dominant role in telecommunications. An important step has already been taken, in that a standard for a Pan-European system was set in Copenhagen in September 1987 by the principal telecommunication administrations in Europe.

Data Communications

The pace of change in information systems is increasing in all areas. All significant trends—expanding customer needs, the proliferation of workstations, and the globalization of the business environment—point to a need for increased communication facilities and integration of various systems.

TABLE 5 Number of Cellular Telephones per 100 Inhabitants

Country or Region	Phones/100 People
Scandinavia	1.4
Austria	0.2
United Kingdom	0.15
Japan	<0.14
United States	<0.14
Rest of Europe	<0.1

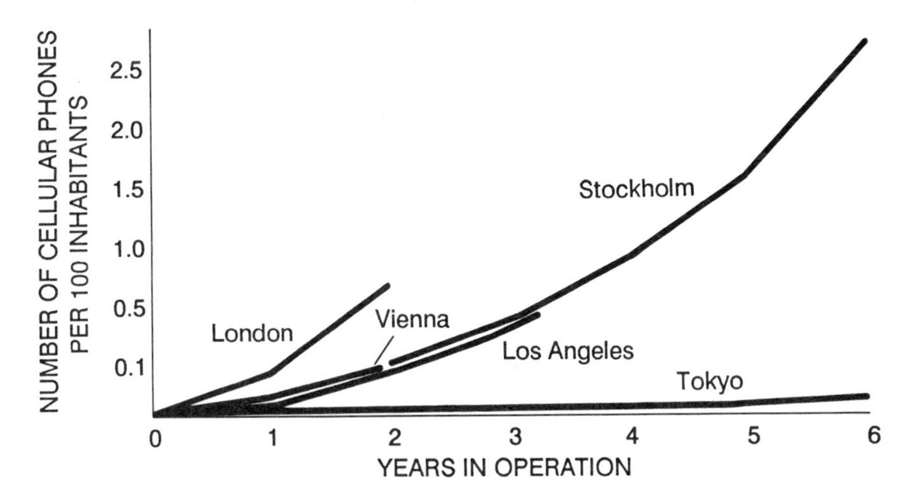

FIGURE 13 Use of cellular telephones (number per 100 inhabitants) in several major cities.

Traditionally, information systems have been used to reduce administrative costs. In recent years, however, this role has been changing as leading companies adopt new strategies with information systems as central components. To carry out such strategies, the information systems must handle both technical and economic information from many systems at many geographical locations, all in real time.

The use of screen-based workstations among white-collar workers has increased dramatically over the past 5 years. Today there are more than 15 million workstations in use among the 60 million white-collar workers in the United States (see Figure 14). By 1990 the number of workstations is expected to grow to at least 35 million. The majority of these workstations are data terminals or personal computers, although the line between these categories is blurring. It is clear that in large and medium-size organizations, almost all of these workstations will be communicating with each other.

The developments in VLSI and software technologies have made workstations considerably less expensive. In 1979 the cost of a personal computer was 25 percent of the total salary of an office worker, but by 1987 this cost will be down to about 6 percent (see Figure 15). Workstations, like telephones, are now considered necessary tools for the workplace.

SOCIETAL IMPACT OF INFORMATION TECHNOLOGIES

Cutting-edge technologies have driven the development of information technologies, which in turn have driven the development of society. This process can be seen in Sweden's development from 1880–1990 as it evolved

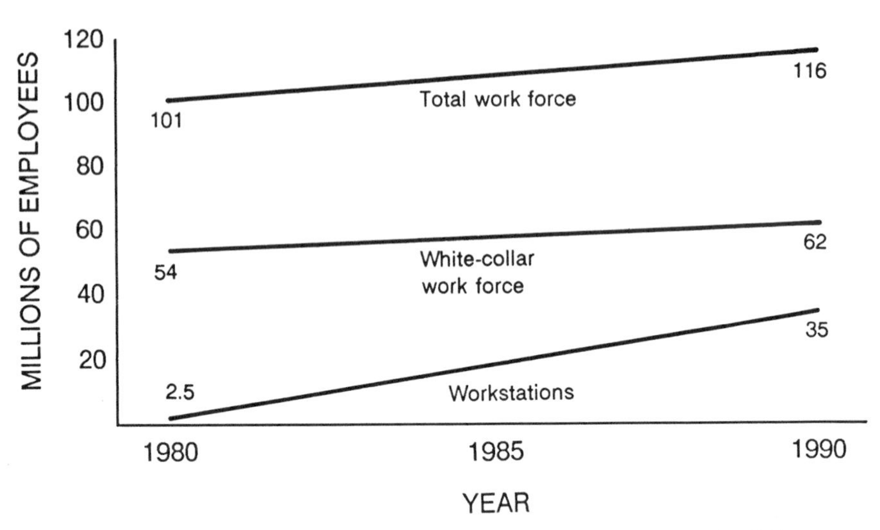

FIGURE 14 Projected workstation penetration of U.S. white-collar workers in 1990.

from agriculture and manufacturing into an information society. The entire Western World has experienced similar development. An information society, with guiding values of quality and the meaning of life and human relations, presents us with many new opportunities. The only limiting factors are the need for more networking and cooperation among organizations, small teams, and individuals.

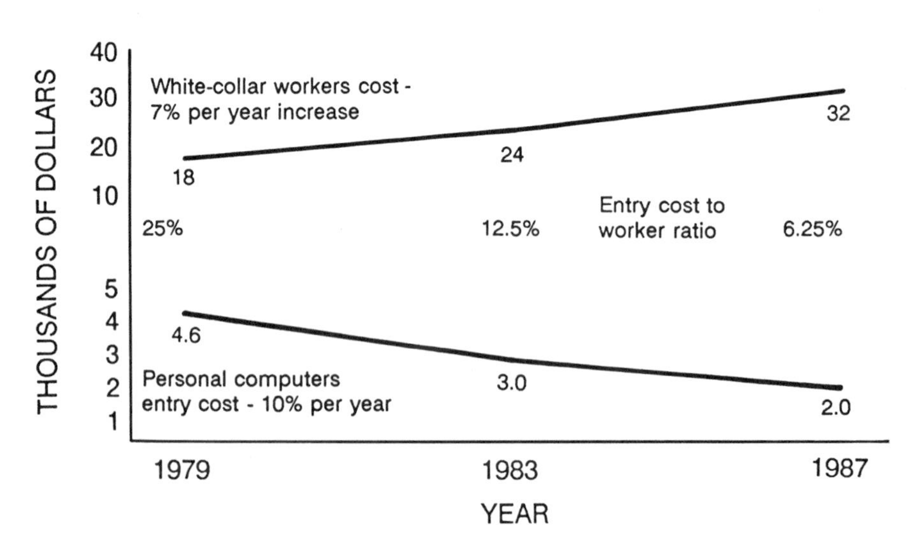

FIGURE 15 Personal computing entry cost. SOURCE: Gartner Group, Inc., Stamford, Conn.

With the new information technologies at hand, we will together be able to form a society that contradicts the frightening visions of the future described in George Orwell's *1984* or Aldous Huxley's *Brave New World*. Communication implies the dissemination of information, and thus of understanding, which will be the basis for democracy and peaceful development in the future.

Technological Advances and Challenges in the Telecommunications Sector

HIROSHI INOSE

REMARKABLE PROGRESS in telecommunications technology has had, and will continue to have, an enormous impact on telecommunications manufacturing and service industries. In particular, digital technology that integrates transmission, switching, processing, and retrieval of information provides opportunities to merge various service modes into an integrated whole. This digitalization, merging the communications and computation functions, has been made possible by dramatic advances in device and material technology, including integrated circuits and optical fibers. As the role of digital processing increases, systems and services become more intelligent and labor-saving on the one hand, and more software-intensive on the other.

Satellites and optical fibers, among other technologies, contribute significantly to the globalization of telecommunications services. Standardization and interoperability of systems have become global issues, as have compatibility of regulatory measures that ensure free trade in telecommunication products and services.

Because telecommunications are now indispensable to socioeconomic activities, reliability and security of telecommunications services have emerged as central issues. In our information age, information retrieval is gaining in importance, while concerns are surfacing about the integrity and authenticity of the information to be provided, as well as the protection of privacy. These diverse issues are important to the future of telecommunications industries.*

CONVERGENCE OF SERVICE MODES

Rapid innovation in information technology has made a variety of traditionally separate information services increasingly related. This trend is often

*For further discussion of these issues, see H. Inose and J. R. Pierce, *Information Technology and Civilization* (New York: W. H. Freeman, 1984).

referred to as the convergence of service modes; the result is a drastic change for telecommunications products and services.

For example, telecommunication has already merged with information processing to provide data communication or on-line processing. Facsimile communication service provided by common carriers and electronic mail service provided by the post office will soon merge, eliminating the physical delivery of documents to and from customers. The difference between videotex by telecommunications services and teletext by broadcasting services will be minimized when cable television systems acquire two-way capability.

One benefit of the convergence of service modes is that it provides economies of scale; that is, many kinds of information can be provided in various forms through a variety of media at a reasonable cost. These benefits, however, will be lost without a reevaluation of regulatory measures, which traditionally have been organized on the basis of separate services. Because services cover broad areas that transcend national boundaries, international compatibility of these regulatory measures is necessary to ensure the unobstructed flow of information globally.

MICROELECTRONICS REVOLUTION

The invention of transistors and the subsequent progress of solid-state circuit technology revolutionized information technology, bringing such innovations as digital transmission, digital switching, and digital computers. The advent of optical fibers, lasers, photodiodes, and other photonic devices permitted lightwave communication over great distances. Clearly, major systems in modern telecommunication have been deeply dependent on innovations in the area of electronic devices and materials. This trend, often referred to as the microelectronics revolution, will intensify in the years ahead. New systems will depend more on the development of new devices and materials.

Characteristically, the technology supporting electronic devices and materials is capital intensive and quickly obsolete. Year after year, billions of dollars have been spent worldwide for research, development, and production of sophisticated devices, which, through keen competition in the marketplace, became cheaper and cheaper and eventually obsolete. Because of the rapid progress of large-scale integration that permits a sizable system to be mounted on a single chip, system manufacturers must also face the issue of capital-intensive investment and quick obsolescence.

SOFTWARE CRISIS

As telecommunication technology becomes increasingly digital and computer-oriented, one major problem is the rising cost of software development and production. Although advances in device technology are lowering hardware costs, software costs are soaring. The increase is due to the constant demand for more sophisticated and diversified types of software, as well as to the high labor costs associated with software development and production.

Several techniques, including structured programming, yield considerable improvement in software productivity. Yet much progress is needed before we will achieve dramatic improvements in software development, production, and testing. Because microprocessors are being used in an increasing variety of applications, tremendous efforts will be required to produce the many programs needed for specific applications.

Another problem associated with software is patent and copyright protection. Generally, patent protection has been given to hardware-oriented inventions. However, since algorithms are considered similar to mathematical formulas or laws of nature, it has been ruled that software-oriented inventions are not patentable. Some countries have amended their copyright law to allow specific programs to be copyrighted. However, in general, copyright protection cannot prevent infringement, as shown by numerous cases in which copyrighted articles are pirated.

STRUCTURAL CHANGES IN INDUSTRY

Whenever industry has changed in structure, workers in traditional industrial sectors have become obsolete and lost their jobs, and newly emerging industrial sectors have suffered from a shortage of workers. The present structural change brought about largely by information technology is no exception. Skilled workers, such as those who assemble telecommunication equipment, are losing their jobs to large-scale integration and the increasing use of industrial robots. Fewer telephone and telegraph operators and maintenance crews in switching centers are needed because of automation and digitalization. Jobs of general office workers are also threatened by the rapid penetration of word processors and other office automation equipment.

On the other hand, many job opportunities are being created in the area of software production. Extensive education and training will facilitate a smooth shift of the labor force from decaying to emerging areas. Such preparation is crucial in dealing with the enormous and unprecedented change now under way in the industrial structure.

Production of telecommunications software should be shared through an international division of labor. Because the needs of users differ from country to country, programs for specific applications have to be produced locally. Developing countries, where wages are relatively low and job opportunities for educated people are insufficient, may have an advantage over some developed countries, where wages are high and people are not motivated to adapt to structural changes in industry. In fact, quite a few newly industrialized countries have been active and successful in the area of software production. Thus, information technology, through its impact on industrial structure, may provide the impetus for a country's movement from developing to developed nation status. By the same token, developed countries that do not recognize and act on these trends will be unable to compete in this area.

STANDARDIZATION

As telecommunication services become global in scale, issues of standardization and of maintaining interoperability between systems and equipment have become extremely important. Standardization, however, is difficult to achieve for several reasons. First, the pace of innovation in telecommunication technology is very fast. Standards set too early may jeopardize future innovation, and if set too late, they are never used.

Second, because of networking, telecommunications require extensive standardization in software and hardware. A large amount of software at each switching center has to be standardized to attain economies of scale, ease of maintenance, and interoffice signaling. To enable communication between terminals and computers, sophisticated software standards known as communication protocols have to be developed and implemented. As equipment becomes more sophisticated, more software has to be standardized.

Third, because new products are designed by competing industries, their specifications tend to be diverse. This makes it particularly difficult to establish a single standard in such areas as computers, terminals, and video packages, where powerful market forces and emerging technologies make existing products and specifications obsolete.

Fourth, conflicts of interest may occur between countries or groups of countries in trying to establish a single standard, because such a standard would benefit some countries more than others. Hence, questions of fairness and political considerations must be addressed along with the technical issues.

Standardization in telecommunication technology requires a great deal of collaboration and compromise between governments, common carriers, and manufacturers. It also requires a thorough understanding of the state of the art, as well as insight into the future activities of all participants, not only governments, common carriers, and manufacturers, but also academicians and user representatives.

If it is impossible to establish a single standard, compatibility or interoperability between standards should be maintained to make interconnection possible. The open system interconnection (OSI) is typical of efforts to ensure interoperability between computers and terminals of different makes and models.

One encouraging note is that, despite two world wars and other international conflicts, worldwide standardization activities have continued and even accelerated in recent years. Standardization is an important area of international activity in which countries, industries, and individuals of diverse backgrounds and interests can think and act constructively and cooperatively for the good of humanity.

RELIABILITY AND SECURITY

Telecommunication systems are subject to a variety of external and internal disturbances. External disturbances include physical factors such as electrical noise, powerline failure, and natural disasters as well as human factors, such as operators' mistakes, vandalism, and unauthorized access by outsiders. Internal disturbances range from chance and wear-out failures of components to hardware and software design errors not detected by testing.

Extensive studies of physical phenomena contributing to component failure have led to better component structure and fabrication techniques. Progress in large-scale integrated circuit technology under strict quality control has drastically reduced electronic device failure. Progress in redundancy techniques now makes it possible for a system to continue operating even when some of its subsystems fail. Automatic diagnosis and plug-in repair techniques have reduced the repair time for complex systems remarkably.

As hardware becomes more and more reliable, attention has been focused on software reliability. In most cases, software failure is caused by some imperfection that was not detected at the testing or debugging stage. These imperfections range from simple coding errors to complex mistakes or misconceptions in software design. Although various techniques have been developed to avoid errors in design and coding, making testing and debugging easier and more nearly complete, software is still less reliable than hardware because it is produced predominantly by humans, who make mistakes more often than machines do.

Security has also been addressed through various techniques that provide secure telecommunication services. Although switching centers and computers are protected by various means from unauthorized access, fire, and some natural disasters, telecommunication systems are still vulnerable to theft. For instance, a microwave link can be intercepted by a highly sensitive receiver from a distance of a few kilometers with a low probability of detection, and a satellite link can be intercepted anywhere. A magnetic disk pack or magnetic tape can be stolen or copied easily. For enhancing security, encryption is an indispensable tool. Various encryption techniques have emerged ranging from such practical methods as the Data Encryption Standard to highly sophisticated public key systems. Although some of the cryptographic systems would require an enormous amount of computing time to crack, they may not confer perfect protection, given the speed with which supercomputers are changing. Enhancement of liability coverage and backup safeguards is necessary to complement these technological measures.

Integrity of Information and Protection of Privacy

Information comes to us from diverse sources, even when it is supplied through a single telecommunication network. It has become extremely difficult for users, who are increasingly dependent on information, to know the original source of the information. Therefore, information providers should expend a great deal of care in gathering and handling data to maintain the integrity and authenticity of the information and make certain that users can determine the source.

This goal can be achieved in a number of ways. Full documentation of sources and methods is essential. Only authorized persons should be allowed to enter important data into data bases, to change data, or to process data for use by others, and there must be a record of these people and their activities. An audit trail must be provided through which one can trace entries into, and changes in, data bases and all steps in processing.

When information concerns individuals' or organizations' backgrounds and knowledge of activities, integrity of information implies confidentiality. If the information gathered is private, proprietary, or confidential, its disclosure might damage a person's reputation or be financially injurious to an organization. Such information should be gathered only for the most compelling reasons, kept only as long as necessary, and guarded diligently against illicit use.

CONCLUDING REMARKS

These technological advances represent only a few of the recent developments that have had a tremendous impact on telecommunication industries. In light of increasing dependence on information, our society needs an enormous stock of information, as well as appropriate means for selective access. In other words, the use of a variety of data bases and the development of data base management technology will significantly influence the growth of telecommunication industries. Because the ultimate objective of telecommunication is the promotion of mutual understanding and the enrichment of culture worldwide, technology that reduces language barriers, promotes computer literacy, and enhances computer-oriented music and arts will expand the horizons of telecommunication industries. All of these and other opportunities should be taken into account in assessing future changes in telecommunication industries.

Technological Advances in the Construction Sector

ALDEN P. YATES

IN THE EUROPE OF THE MIDDLE AGES, craftsmen with varying levels of skill manufactured carts one at a time. Even the best of these carts provided little more than basic transportation. Yet in the same cities of Europe, at the same time, master masons and builders created incredible stone cathedrals, using principles of design and construction that were breathtaking for the time.

Today the technologies of both manufacturing and construction have changed, but not nearly to the same degree. Part of the reason for the different degrees of change can be found in the basic differences between manufacturing and construction. Construction is essentially the process of moving and assembling materials and equipment into a completed, operational facility. Although many construction operations are repetitive, they are performed neither in a fixed sequence nor at a fixed location. Also, since construction, unlike manufacturing, rarely involves production of a standardized product, the demands on the material supply functions of buying, expediting, receiving, warehousing, and delivery are much more complex. For many of these reasons, the basic construction process of building stick by stick, piece by piece, has remained unchanged since the Middle Ages.

But that basic process is critical to the world's economy. Construction is larger than any single manufacturing segment of the U.S. economy. It contributed $174 billion, or 4.7 percent of the gross national product (GNP), in 1986, whereas all manufacturing contributed 22 percent of the GNP. Residential work was 47 percent of the total contract awards. Commercial work awards were 19 percent. Heavy construction—which includes utilities, pipelines, and other energy work—was 17 percent. The picture is equally impressive beyond our borders. For example, in 1984, the value of new construction put in place, defined as new residential and nonresidential construction but generally excluding maintenance and repair construction, was $317.2 billion for the Soviet Union, $200.1 billion for Japan, $71.3 billion for the Federal Republic of Germany, and $43 billion for the United Kingdom.

The Great Wall of China, the space shuttle launch facility, a petrochemical plant, a neighborhood shopping center, the Erie Canal, a nuclear power plant, a single-family home—all are construction projects, yet each requires different skills and technologies. Collectively, they represent the many sectors of the construction industry.

The residential and commercial construction sectors involve the creation of facilities that are essentially structural in function. These facilities include the service utility systems necessary to support the people who use them, including power distribution, heating, ventilation, and lighting. By contrast, the industrial sector creates facilities incorporating industrial process systems and equipment designed to produce an end product, such as automobiles, textiles, chemicals, refined metals, or electric power. The heavy civil sector encompasses major public works, including dams, highways, airports, and water distribution and sewage facilities—in short, most of what we now call infrastructure.

Over the past 10 years the impacts of technology on the construction sector have varied by the type of construction being performed, but in general, the changes have been largely evolutionary. Today's constructors have not come as far from the cathedral builders of the Middle Ages as today's automakers have from the cartwrights. In the future, however, there is a high potential for significant developments that will change the basic nature of construction. These developments will capitalize on advances already apparent in other sectors. They will be global in origin and in scope, with applications driven by both continued technological innovation and competitive pressures. They will include direct technological impacts on the performance of specific construction activities and major changes in the manner of managing a construction business.

This paper addresses the most significant changes in construction by examining technological trends and how they affect the entire construction sector. These trends fall into four major areas: construction-related design; construction equipment and methods; automation and expert systems; and construction management.

CONSTRUCTION-RELATED DESIGN

Computer-aided design, or CAD, is now a fact of life in the design-construction process. The benefits to the construction industry already have been significant in several respects. These include reduced interferences, which are instances where the design of separate systems, such as electrical conduit and high-voltage alternating current ducting, compete for the same physical space; better feedback to the design cycle on the impact of constructability enhancements, which are design factors intended to simplify construction and reduce job hours; and improved communication between the designer and supplier to ensure that the right component is available at the construction site when it is needed.

CAD also has been a positive adjunct to the process of "fast-track" construction, an approach in which engineering and construction proceed concurrently. In this approach, construction might begin when 40 percent of the project's design is completed, rather than waiting for the design to be 100 percent finished. CAD systems, which are used extensively in commercial and heavy industrial construction but have yet to prove cost-effective in residential work, are able to generate design information faster and more accurately and can implement midstream changes with more ease than conventional drafting can. Construction work is able to begin earlier because the design is more rapidly developed. The savings in time-related design costs can be significant; in the time it takes an engineer at a drafting table to produce one drawing, an experienced CAD technician can produce four.

There are several ways to classify major CAD trends, including the trend toward engineering workstations as special-purpose computer terminals dedicated to the automated design process. Such standard design details as typical civil, structural, piping, and electrical schematics are now available on many CAD systems. The trend toward lower-cost, more powerful workstations will continue indefinitely, although the cost per unit will probably stabilize. Further, most design depictions will be achieved by means of three-dimensional computer models with sophisticated, standardized design symbols and aids. There will be a reduced need for today's orthographic and isometric drawings when three-dimensional models are available.

With the rise in CAD has come the need for common CAD products. Compatibility is still a problem, because vendors continue to emphasize enhanced features on their own systems, as opposed to compatibility with other systems. Some standardization efforts have made considerable progress but require further efforts to be fully effective. For example, in the area of initial graphics exchange specification (IGES), a loosely organized group of CAD vendors, users, and manufacturers has recently undertaken a collaborative effort to develop IGES guidelines for standardized design symbols and common design standards for all CAD machines.

Improvements are continuing in checking the interfaces between standard systems, such as between mechanical and structural systems or between electrical and plumbing systems, including the use of expert systems in limited ways. Closer coordination at these interfaces reduces the risk of overruns in construction cost and increases the reliability of the construction process.

One example of the use of expert systems developed in the Bechtel Group is a three-dimensional coordinate system that allows people to "walk through" a facility before it is built. It significantly benefits construction by reducing interferences because the facility and its systems are displayed in three dimensions, not just the two dimensions of traditional plan and section drawings. The system automatically raises a flag when two components in the drawings occupy the same physical space, and allows corrections to be made before the problem reaches the field. Future improvements in design capabilities will allow closer linkage of design to operations and maintenance, better life-cycle costing linked to design alternatives, and improved methodologies for cost estimating and procurement linked to electronic design program software.

These enhanced design capabilities allow earlier review of all designs from a constructability viewpoint, benefiting all members of the project team. Further, linkage of design documents to computerized simulation of the built facility is gradually becoming common, optimizing design from an operational viewpoint. These trends benefit the owner, who will push hard to include them as a standard part of the design process.

The drive toward CAD has implications beyond the improvement of the design process. For example, electronic communication of data among the owner, designer, material supplier, and builder of a project is a very significant trend, so the industry must prepare for all project team members to have common networked workstations and to meet electronically. Bar-code technology is now playing a role in tracking and locating materials and equipment, with identification codes linked directly to CAD systems.

CONSTRUCTION EQUIPMENT AND METHODS

Construction equipment, in general, assists in moving and assembling materials. Emphasis has been on moving larger pieces or on moving material faster, with greater reliability and accuracy. In the recent past, improvements in such conventional construction practices as slipforming—the use of a moving form for pouring concrete—have continued on an incremental basis. Heavy equipment for use at the job site, such as cranes, conveyors, and earth movers, continues to become more efficient. Dramatic improvements have been made on specific machinery, such as laser-based survey equipment, laser-guided excavation equipment, and new tunneling equipment.

Today's job site also features the more prominent use of advanced materials: honeycomb structures and foams for greater strength; polyester fiber for improved durability in the refitting of sewage and water pipes; fiberglass fabric for rapid repair work; and specialized materials for arid, arctic, undersea, radioactive, and extraterrestrial environments. However, the basic building blocks of construction—steel and concrete—are expected to remain relatively unchanged. Specific qualities of these materials will be improved, but no major substitute material is on the horizon. All types of construction are being affected by these trends in construction equipment, methods, and materials, although customer preferences are a significant restraint in some areas. The preference for wood and natural materials over plastics in residential construction is a good example of this restraint.

Off-site fabrication and assembly is a trend that has been clearly established. Equipment modules of 2,500 tons are no longer uncommon, and the trend toward even larger modules will continue. For example, lifts of 10,000 tons have already been made in the construction of North Sea oil platforms. Although certain constraints still exist, such as proximity to water transport and the limits of lifting equipment, the pattern of maximum off-site assembly and close coordination of delivery logistics is expanding.

Factory assembly of components has several inherent advantages over job site assembly. For example, a heat exchanger once assembled on-site piece by piece is today fabricated in a vendor's shop on a structural steel skid complete with ladders, railings, wiring, piping, and instruments and is shipped essentially ready to plug into other modules on the site. Assembly performed in the controlled environment of the fabrication shop has distinct advantages: It avoids the climatic extremes of a field site and benefits from better management of material and parts inventories, maximizing productivity and quality. Similar shop fabrication processes also are used for instrument panels, compressors, pump units, and switch gear buildings. Schedule time is saved, because more assembly work can be done in parallel with on-site activity. Trade interferences are tested and resolved earlier. ''As-built'' drawing documentation is minimized, thus reducing the potential for design errors or omissions. Labor costs are reduced because (1) the controlled shop environment permits increased automation, which enhances quality and productivity; and (2) shop-performed assembly generally has lower wage rates than comparable work done by field construction personnel.

The construction site is characterized by a high level of activity as men and equipment move materials, tools, and design information from one place to another. A certain degree of inefficiency is normal, depending on the size and complexity of the project, the constraints of procedures, and the degree and effectiveness of detail planning and scheduling. As the use of modular and prefabricated construction methods increases, this inefficiency will decrease. This will hasten the use of robots, which will reduce cost and improve safety; increase productivity, which will shorten construction schedules; and allow for automatic on-line inspection, which will yield a higher-quality product.

Modularization applications, to some extent, will follow the path from simple to complex components. Assembly of similar warehouse and low-rise office buildings is now fairly common. Assembly of similar structural components and fixtures for more complex facilities, such as hospitals and high-rise office buildings, is also part of this trend. In addition, individual heavy components, such as large power-generation boilers, are beginning to be manufactured in the factory and then disassembled, shipped separately, and reconnected in the field using advanced manufacturing techniques.

However, not all technological trends point toward fabrication away from the job site. For example, the automation of field welds by means of standardized robotic devices will provide reliability at the job site as high as that in the factory, and may even be more cost-effective for those materials that are best shipped in smaller pieces. We must be careful to balance our assessment, since advances are taking place on both sides of the fabricator–job site equation.

AUTOMATION AND EXPERT SYSTEMS

The use of computerized expert systems for construction applications is a growing trend. Current examples include systems to diagnose vibration problems in rotating machinery and systems to verify weld performance qualifications. Extensive research to develop construction-based systems is under way at both the U.S. National Bureau of Standards and the U.S. Army Corps of Engineers in such areas as evaluation of concrete durability and building air infiltration dynamics. The use of expert systems will probably be the most important application of artificial intelligence techniques for construction over the next decade. By the turn of the century, there is good potential for increased use of self-directed robots controlled by expert systems. Such advanced-application robots would finish concrete and spray paint buildings (already being done in Japan), apply sprayed insulation to structural steel members, and even install structural steel. Robots in construction would differ from those in a manufacturing or production line setting, where the robotic units generally are stationary and tasks are performed on products as they move by. In construction, the building is stationary and the robot would have the ability to move about in the performance of its tasks.

Technologies such as laser range-finding and geodetic positioning can be used to pinpoint exact locations, to automate storage areas on the job site, and to set guide tracks for remotely operated vehicles. These technologies will gradually be integrated into a coherent system for the highly automated control of certain job site activities.

Automation in the construction sector is usually seen in terms of robotics, and the development and application of robotic systems in all industry sectors is relatively new. According to the *EPRI Journal**, only 6,000 robots were delivered in the United States by the end of 1981, and most of those were

*Moore, T. 1984. Robots join the nuclear workforce. EPRI Journal 9(9):6–17.

installed in the preceding 7 years. According to Robotic Industries Association estimates, the total number of robots installed in the United States rose to more than 20,000 at the end of 1985, up from 14,500 at the end of 1984. Some experts predict that as many as 100,000 robots may be at work in this country by 1990, with more than 1 million projected worldwide. These are classic robots—programmed, repetitive machines such as those that are used in production line operations. However, the technology is directly transferable to remotely controlled robots more applicable to the construction site, and it is expected that the development and application of such robots will parallel the expansion of production line robots. First-generation construction robots now on job sites are really microprocessor controllers retrofitted to conventional construction equipment. The use of remote technology is accelerating in areas where laborers are performing repetitive tasks or working in a hazardous environment, or where quality can be improved by continuous inspection of the operation or product.

For example, at the Three Mile Island (TMI) nuclear facility in Pennsylvania, certain areas of Unit 2 are too highly contaminated for workers to enter and perform cleanup activities. As a result, robots—or more precisely, remotely controlled vehicles—have been used instead of humans to carry out decontamination assignments such as surveillance, washing of walls, removing radioactive materials from concrete surfaces, and suctioning sediment from the floor. "Rover," the robot used at TMI, was developed by Bechtel engineers working with GPU Nuclear, Carnegie Mellon University, and the Electric Power Research Institute. It is a six-wheel-drive robot that can negotiate turns and climb over obstacles. It is operated by tether from a control room.

Feedback of as-built conditions is the step that closes the design-construction loop and will enhance robotic applications in construction. In this process, erection data from construction are fed back in real time and compared with the design data. Installations that exceed tolerances can then be corrected. Of critical significance in the field is the accumulation of deviations; although each of these deviations may be individually within construction tolerances, together they may be enough to defeat automated equipment. The feedback of as-built dimensions can eliminate this problem, as it is usually not necessary to construct to exacting tolerances, but only to know exactly where the construction is located. With feedback, succeeding elements can be adjusted to fit.

The use of robotics for construction, operations, and maintenance or design engineering activities has applications worldwide. For example, a Japanese company has used an articulated robot to weld small-diameter pipe. As a result of the high quality of the weld, the finishing process became unnecessary, and plans are being made to automate all processes by combining the robotized system with computer-aided design and computer-aided manufacturing (CAD/CAM). A Finnish firm is developing a gantry-type welding robot capable of automatically joining ship sections up to 15 meters long. The robot is able to learn from its mistakes and make the needed adjustments. Other firms have developed an approach to the robotization of painting and blasting and the development of wall-walking robots that use vacuum pads or magnets.

CONSTRUCTION MANAGEMENT

The construction management process—the system of controls that optimize the design, procurement, and construction process—is key to the ability of the construction industry to capitalize on technological innovations. The process of planning, scheduling, and cost control must address the interfaces between all disciplines and provide the framework that allows new technological developments to be assimilated efficiently into a construction project.

From the management perspective, such technological trends as the increased use of job site robots, with their high capital costs and 24-hour availability, demand improved just-in-time delivery systems for precise material scheduling. This, in turn, requires sophisticated computer data bases linked to design, purchasing, and warehousing systems to ensure effective management. In this sense, most large-scale construction work is becoming essentially fast tracked. In the fast-track approach, design and construction proceed simultaneously rather than waiting for engineering to be completed in advance of construction. This technique is intended to require less time for engineering and construction than is required in a conventional one-step-at-a-time approach. Fast-track construction can be used on any size project and refers to speed of completion rather than size. In addition to these general trends, specific evidence of the impact of technology on the construction management process can be cited, particularly in the areas of bulk staging activities, inventory, construction start-up, training, quality control, and information handling.

Earlier identification of bulk component requirements, allowing earlier bulk staging of commodities at predescribed locations, is a consequence of rapid design capability. We are essentially moving toward an assembly-line process for even the most complex construction sequences.

The application of a modified just-in-time method of supply to construction sites can be used as a method to guide job site activity. In this concept, minimum warehouse and lay-down areas are used, and only small items are inventoried. However, since the contract risks due to faulty delivery systems at the construction site are at least as onerous as those at the factory, we need to adapt factory-based just-in-time delivery schemes to the construction environment.

In job sites of the past, the first tools unpacked were the shovels. Today, the first things out of the box are often computers. Terminals and graphic workstations give access to the released design documentation. Paper plotting, to the degree required, will be done at the job site. These computers provide communications links with the engineering office, the client, and relevant vendors or fabricators. Over time, the establishment of computer information networks will become a key start-up activity at construction sites, since it will be a crucial part of the construction management process.

The size of the skilled labor force required at the job site will continue to shrink as machines are used for more complicated tasks and to help manage site logistics. However, the people still needed will have to have higher skill levels to operate more complex devices. The required training process will place significant demands on the construction management staff.

The impacts of continued technological change on quality control practices in the construction sector must also be considered. For example, with more accurate robotic machines, we may see dramatic changes in two areas: reduced need to check certain construction devices that already have imbedded high reliability, and the availability of machines that improve the quality-checking process itself. All of these developments reinforce the emphasis the constructor must place on the reliability of the completed construction project. As capital costs rise along with such production costs as labor, raw materials, and energy, increased emphasis is placed on plant availability, a measure of the project's reliability.

From an administrative standpoint, the proper management of change orders for a job already under way has always been a difficult task for the contractor, with often frustrating results for the entire project team. Improvements in information handling will include the gradual automation of much of the change order process, resulting in smoother activity on the project, savings in time and money, and added flexibility for the owner of the facility.

All of these developments point to the need for construction management personnel—especially site managers—who are able to marshal new technology and apply it effectively in an environment that historically has been less than progressive. Complex issues such as trade union practices and the inertia of field engineers and superintendents must be addressed. Restrictive craft union work rules and jurisdictional disputes have contributed to the steady advancement of the nonunion movement. The application of robots, advanced modularization, and other new construction techniques and methods will be successful only if craft labor and its leadership make a quantum jump in acceptance of new technology. Similarly, site management must plan the transition carefully to ensure that a positive "make-it-happen" attitude exists among both craft and supervisory staffs.

If these new techniques are to be effective, they will require a level of organization and discipline seldom seen on a construction site. Thus, management must be skilled both in the assimilation of new technology and in the handling of the human factors issues arising from this technology. The challenge is significant and will require training in disparate fields including production control and human relations. The task is manageable, but past problems with the adaptation of limited technological advancements in the construction sector suggest that this issue may be the most difficult of all.

THE IMPACT OF TECHNOLOGY ON CONSTRUCTION MARKETS

Since before the building of the Pyramids, a fundamental business relationship has existed in the construction sector among the owner, the designer, the material supplier, and the builder. Technological change cannot destroy this relationship, but it will continue to cause subtle alterations in the patterns that characterize it. In particular, the increased capability to catalog and manipulate knowledge will enhance the role of those owners who become more involved in the study of options available to them. Further, the global availability of basic design–build capability will permit tighter competition on a technical level for conventional construction projects. The so-called first-of-a-kind projects involving sophisticated custom design and construction will still require specialized capabilities, and firms will meet these requirements using advanced tools and techniques. In the widest sense, construction itself will feel the impact from several continuing global trends: (1) a shift toward more decentralized market activity with more players, both large and small; and (2) increasing access throughout the world to more information about basic human needs such as food storage and sanitation, particularly in developing countries. The point here is that as knowledge is disseminated more broadly and deeply into global society, there will be increasing understanding of both the need for facilities that satisfy basic human needs and the knowledge of how these facilities can be built simply and cheaply.

Technology is changing the nature and shape of the markets served by the construction sector, just as it is changing specific activities in the construction process. One of the most obvious of these changes is the addition of new markets because of technological progress in general. Two examples are the so-called clean rooms required for the assembly of complex electronic components, and the gradual expansion of construction work related to the containment and disposal of toxic wastes. The design and construction of facilities for innovative sources of energy is an ongoing challenge. Extraterrestrial activity—construction in outer space—motivates unusual solutions and creates new possibilities. The more prosaic but no less important requirements of infrastructure replacement, such as underground urban utilities, call for extensive application of new construction capabilities to minimize costs and disruptions. Further, to the extent that technology allows a wider range of design options to meet market needs, the construction sector will respond with new interactions with clients and diversification into new business lines.

Technology is also continuing to change the way we create and modify building codes. In the era before robotics and automation, code-setting required attention to a considerable number of safety factors based on the uncertainties allowed for in many design calculations. Despite the traditional constraints that retard the changing of codes, it is likely that various automated construction technologies will reduce physical quantity requirements and costs considerably, simply by reducing the overprotective limits in some of today's codes.

FINAL OBSERVATIONS

In the next 10 years, the greatest technical impact in the construction sector is expected to come from improved management methods and automation. Advancements in management methods to improve productivity and schedule performance will employ automation and expert systems to a great degree. Construction design will see increased sophistication in the conceptual phase and real-time data base communications networks to support estimating, scheduling, and project management.

Procurement activities will improve by the use of CAD systems to provide a direct interface with major vendors and suppliers. Automated warehouses, staging areas, and related support facilities will also play key roles. In the construction process itself, expanded use of computerized scheduling, tracking, and control using real-time networks and robotics-assisted operations will play increasing roles, where practical, to meet quality, safety, and cost objectives.

Several major companies and government institutions throughout the world have active and comprehensive construction research programs, including companies such as Bechtel and countries such as the Federal Republic of Germany, Sweden, and the United Kingdom. These programs are executed sometimes in company laboratories. The companies are competing in all overseas markets. In Japan, for example, more than 50 construction companies have their own research facilities. Such laboratories are part of parent company activity to expand vertical market positions. Significant efforts are expended worldwide on applied research to develop new construction technologies. Although some firms liberally promote the more glamorous aspects of construction research and development as a marketing tool, many have planned seriously for improvements to their knowledge base. The knowledge gained by this research will undoubtedly improve their competitive edge for major construction jobs. All firms in the construction sector must reexamine the cost-effectiveness of their research and development commitments in relation to the competitive advantage they expect to derive.

In the future, design and construction, in general, will become ever more closely integrated. More effective design optimization studies and constructability decisions will be made during the engineering phase. We will, in effect, maximize the use of automated job site machinery by designing with it in mind. Procurement, as a generic process, will also be more closely tied to design decisions. Trade-offs among procurement, scheduling, and constructability will be more easily understood through improved, computerized analysis of procurement options.

Construction itself will undergo significant changes in methods of management and work performance. Technology is having major impacts on methods and systems for constructing all types of projects. The most significant challenge will be that of management coordination. Historically, the management of construction activity has been too reactive—dominated by archaic methods, restrictive trade union practices, and ineffective planning. In some ways, we are still building the cathedrals of the Middle Ages when we need to build space stations and advanced factories on earth.

The technological capability exists for vast improvement in our methods, and it will be our management effectiveness that determines our success in the long run. With the vision of what can be done and the commitment to make full use of available technology, the future for construction should hold many extraordinary developments.

Fifteen Years of Major Structural Changes in Manufacturing

PEHR GYLLENHAMMAR

THE MANUFACTURING INDUSTRY is still under great pressure worldwide, having gone through a difficult time in the 1970s. In the late 1960s the attitudes toward the industry were negative, particularly among young people. These attitudes continued into the early 1970s, at which time the industry experienced the oil price increase, several recessions, and hardly any corresponding boom in the economy. In addition, the Japanese offensive which began in the 1970s has presented difficulties for many companies in Western Europe and the United States.

As a result, many industries were ailing. Some were subsidized, some were dying, some were restructured, and some succeeded. It has been widely supposed that in the 1980s the manufacturing-based society would give way to a service society, a postindustrial society, and possibly a society for high-tech industry. However, the postindustrial society has not arrived. The service society is here, perhaps, but will not solve our problems, and high-tech society still has to be defined.

NEW PRODUCTS, NEW MANAGEMENT, COMMITTED PEOPLE: MEANS FOR SUCCESS

Today the manufacturing industry is facing a new environment that is a result of the difficulties of the 1970s and the pressures of today. To survive in this new environment, the industry has to be good at almost everything. Today industry has to excel at both product development and production. It has to be in command of the development of distribution systems, that is, logistics. It must be able to identify target groups in the market extremely well. This means an industry must be precise in its message, offer good service, and look upon its total operations as one system. It was not long ago that a manufacturer could boast about a product that invaded the market simply because it was newer or better. A manufacturer could also specialize in marketing and do reasonably well with products that were not exceptional in any way.

In today's tough environment, new technologies have, to a large extent, changed the product. The automotive industry is the largest industry buying and applying a wide range of new technologies. We use new engineering materials such as plastics, aluminum, and alloys and depend heavily on electronics. Sensors are used extensively throughout the products, and we have taken a systems approach to product development. We see the car as a system more than a piece of hardware or an assembly of hardware. The automotive industry also has new tools for product development. Computer-aided design has led to, if not a revolution, much higher efficiency. Volvo recently has designed an engine without using the drawing board at all. So, engineers are supported in their development work by artificial intelligence, simulation, and sophisticated testing equipment.

NEW DEMANDS FROM EMPLOYEES AND CUSTOMERS

The manufacturing industry experiences other, new demands besides competition and the pressures of recession and tough economic conditions. There are new demands from two kinds of people—the employees and the customers. The employees' demands, of course, are for better education and more information. Employees want more meaningful jobs and a bigger say in the development of the corporation. They want to see evidence that they can experience personal development through the job.

Employees look at what might be described as the invisible contract. They give something to the corporation. They do not always feel that their contribution is reciprocated, and they want to see evidence that this invisible contract is maturing into something that is good for them. They are looking not only for monetary remuneration but also for an interesting job and the possibility of receiving training and trying new career opportunities in the company.

Customers, too, are much more sophisticated today than they were in the past. They ask not only for value for their money but also for care. They ask for service and they too want a kind of invisible contract with the supplier. Consequently, people are the key resource in industry, whether they be customers or employees.

NEW TECHNOLOGY MEANS GREAT CHANGES AND OPPORTUNITIES

In the automotive industry, the standard scenario in the late 1960s and the early 1970s was that large-scale production would survive and smaller-scale production would die. This view was partly tied to the technology that was available at that time. The automotive industry was using transfer lines, capital-intensive automation, and mainframe computers. This expensive large-scale technology led to the concept of the world car, which was talked about throughout the 1970s. Manufacturers chose their manufacturing base wherever costs of production seemed to be lowest and where availability of labor was ample. Then they applied this capital-intensive technology to production.

Today we can see that those industries that have survived were not necessarily the large ones and that small industries can also be successful. In part because of the new technology, there is a mixture of both small and large industries. Automated manufacturing technology provides an opportunity for smaller-scale production to become profitable. Robots coupled with new types of numerically controlled machines provide flexibility. Robot gates are replacing transfer lines. Microcomputers assist or replace mainframe computers on many jobs. Light tools are replacing heavy tools. Trucks used in materials handling are superseded by self-propelled electric carriers. Modern information systems allow for delegation and dissemination of information at low cost. This also means that decision making can occur in many different parts of the production system, whereas previously it had to be centralized, both because of the equipment and because of the lack of proper information technology. Together with light and flexible equipment, computer-aided manufacturing makes it easier to work with different layouts. A manufacturer can easily change the layout of a production system that was previously too expensive or difficult to move because of the heavy and capital-intensive equipment.

Because it is easier to manage a small group of people than a large mass of people, small-scale manufacturing has also become economical. It is now possible for a manufacturer to go from a 5,000-employee plant to a 500-employee plant. Years ago there were suspicions that the 500-employee plant would not be able to compete. Today we have evidence that such a plant can compete and in many ways is more efficient.

As a result of these various technological developments, the automotive industry, which in the 1970s was labeled a mature industry, meaning that it was on its way out, is now called high tech. It is very technology intensive.

AUTOMOTIVE INDUSTRY—BIG CONSUMER AS WELL AS PRODUCER OF NEW TECHNOLOGY

The automotive industry is the largest manufacturing enterprise in the world. The tip of the auto industry iceberg—25 corporations making and selling cars and trucks—is a $500-billion-a-year business at 1987 prices. The rest of the iceberg—supply of materials and services, auto fuels, and the aftermarket, such as sales of used vehicles—is six to seven times larger.

This means that the automotive industry, broadly defined on a worldwide scale, is larger than the entire U.S. economy.

In the 1960s and early 1970s, it was fashionable to demonstrate the automotive industry's importance by citing the consumption of the industry's staple materials diet—steel, glass, plastics, and rubber. These figures demonstrated that the 25 million jobs at the tip of the auto industry iceberg support another 6 million jobs directly in manufacturing alone.

A new dimension to the traditional importance of the automotive industry is now becoming apparent. The industry is a major consumer and producer of new technology hardware and software. It is the biggest manufacturing meeting place or crossroads for innovation and application of design, development, process, product, and distribution technologies:

- Half of the world's robot population works for the automotive industry.
- Auto companies (General Motors, Fiat, Volkswagen, Volvo) are among the major producers of automated production hardware and software. Worldwide, the tip-of-the-iceberg auto companies also carry out about 10 percent of the world's aircraft business.
- The auto industry claims a 50 percent share of all of today's installed flexible manufacturing systems.
- General Motors, Ford, and Chrysler are IBM's major U.S. customers after the U.S. Department of Defense.
- In Western Europe, the auto industry is the dominant customer for new materials (special steels, superplastics, and composites).
- In Japan, the annual combined research and development budget of the automotive industry comfortably exceeds the European Economic Community's funding of collaborative research plus the budget of the European Space Agency.

Cars and trucks lead all other manufacturing industries in cost and product utility per pound of weight. This lead is threatened only by the computer business, whose products are—as yet—remote from widespread consumer ownership and use. However, the potential for further technological advance in the automotive industry is vast.

ORGANIZE THE PRODUCTION SO IT DEVELOPS THE EMPLOYEES

Workers were exploited during 50 years of manufacturing development, from Henry Ford's assembly line up to similar assembly lines that have not yet been deserted. Today assembly operations can be carried out differently by eliminating heavy and monotonous tasks.

Some of these less desirable tasks, for example, have been taken over by robots. Volvo, for example, installed its first robot in 1971. That decision had to be made by the chief executive officer because there were no rules to delegate authority for something that was not profitable, and there was no way to calculate a return on the first robot.

Today a new production organization can be designed in such a way that, to a much greater extent than in the past, it corresponds to the way people wish to organize themselves. Earlier, the heavy, capital-intensive technology had to dominate people, who were really the servants of that system and of that technology. With a new, light, flexible technology, a manufacturer can organize people so that they are in command of the technology—a very dramatic change. Volvo has used this change in programs to develop and promote employees. We have sought to give them training not only in new technologies, such as the lighter, more flexible, and also more complicated equipment, but also in a new responsibility to accept a much greater mandate in minding their own workstations. Volvo has used this approach to extend work cycles from 90 seconds up to 1 or 2 hours, to let the employees organize themselves the way they find appropriate and to do things right and exercise their own quality control. The most common work organization today is a team of 3–12 people who are responsible for a well-defined part of the production. Normally, this includes quality inspection, handling of incoming materials, maintenance, planning, and problem solving.

New technologies also provide opportunities to create more flexible work environments. For example, one of the main problems in large-scale manufacturing or any area of manufacturing that involves large numbers of components—such as the automotive industry—is that of materials handling. Here the available technologies include the self-propelled, computer-guided carriers that help to keep the floor clear of equipment because all equipment is mobile. This technology also makes it possible to change the layout of equipment to fit particular work organizations.

What manufacturers see, therefore, based on the new technology, is a freedom they have not had before. They can either use this freedom to contribute to development of their employees or they can continue to organize their plants in a conservative fashion. Manufacturers will find, however, that the new competitive tool is to organize production so that they can develop their workers. The necessity of employee development represents a major task and a new role for management. Not only does it entail a new kind of technical training but it also includes leadership training that was not even considered a possibility for blue-collar workers 10 years ago. To implement a new management philosophy, Volvo started pioneering plant projects in the early 1970s. The best known of these has been the Kalmar plant. These projects became real and visible symbols for people to learn from and have been the source of experience and productive ideas that have been diffused to other parts of the organization.

LACK OF PATIENT CAPITAL LEADS TO INADEQUATE EVALUATION OF POSSIBILITIES

Manufacturers face many problems related to the fact that the manufacturing industry may not have been successful enough, and the availability of new technologies creates a chance to achieve improved growth. It seems obvious in many parts of Western Europe that the financial returns from manufacturing are too small. It seems clear that this is also the case in the United States. Manufacturers have competition from the service sector, where returns often are more handsome, and from the financial services, where the returns of the last 2 or 3 years have been phenomenal, not only for shareholders but also for employees.

The manufacturing industry that enjoyed more positive attitudes after the bad years of the 1970s is again losing its glamour. The risk is that the industry is subjected to short-term evaluations of its possibilities. Training people to use the technology at their disposal requires time. Industry also needs patient capital. Even if product development cycles are shortened and even if products can be introduced into production faster today than they could 10 years ago, it is still a matter of 5- to 10-year perspectives. It may well be 10 years before a manufacturer is certain of a fair return on a new product development that has been introduced into production and then into the market. Serious investors understand that technology and product development takes time and requires research and development by competent people. Short-term speculations are detrimental to the manufacturing industry. I would predict that the capital market will understand this very soon.

The manufacturing industry is running out of patient capital, which is hardly available anymore. Not only does industry need to apply the new technological means at its disposal, it also needs support to get more glamour, to get more real development. At present the acquisition of industries is often more profitable than adding value through patient and good work.

NO SERVICE SECTOR WITHOUT A VIABLE MANUFACTURING INDUSTRY

It is often said that the service sector is growing faster than the industrial sector, and of course, that is true. But there is a link between services and industry. A great part of the new service- and knowledge-based industry is directly or indirectly a result of the manufacturing industry. For instance, manufacturing companies need more technical support and systems development. Many companies prefer to buy these services rather than build up their own in-house resources. If we take industry or industrial growth away, eventually part of the service sector will collapse. Although the manufacturing industry cannot solve the problems of unemployment and slow growth, even with the new technology now available, it is absolutely certain that those problems cannot be solved without a viable manufacturing industry.

Globalization of Industry Through Production Sharing

EMILIO CARRILLO GAMBOA

GLOBALIZATION IS A KEY TREND in the business world today. The evolution of supply, demand, and environmental factors is driving companies toward operating as if a homogeneous worldwide market existed in their industries. Many forces are pushing for globalization. A decade of peace and increasing governmental advocacy of free trade in all the major developed countries has lowered trade barriers and given a renewed impulse to global trade. In the 20-year period from 1970 to 1990, world trade will have more than doubled in importance, from 12 percent of total world production to 27 percent (McKinsey and Company, Inc., 1987).

Since demand has become homogeneous across borders, producers of major consumer goods today can use similar marketing concepts and approaches to reach the entire Western world. Thus, globalization presents both a major threat and a major opportunity, particularly in developed countries. Industries and companies that previously enjoyed relatively safe home markets now find themselves faced with the possibility of new competition from companies that had never attempted to market products in their part of the world. Globalization has made worldwide competitiveness critical for survival. However, the conquest of global markets will be the reward of the most efficient producers.

To achieve worldwide competitiveness, many managers would quickly move any factory anywhere in the world where they could get cheaper or better materials, labor, and vendors, and where laws and governments were more congenial. More and more manufacturers make parts and subassemblies in different areas of the world and then assemble the complete products elsewhere and sell them in global markets. The location of competitive producers changes constantly, and factories move repeatedly to find the most favorable locations.

This paper analyzes the factors that have led developed and developing countries, especially the United States and Mexico, to adopt this system, termed "production sharing." It examines the results of production sharing to date and discusses possible policy options to achieve the best results of this global trend.

EVOLUTION OF PRODUCTION SHARING FROM THE POINT OF VIEW OF THE DEVELOPED COUNTRY: THE U.S. CASE

Production Sharing and Industry Life Cycles

In discussions of the changing international division of labor, manufacturing industries are often classified as either traditional industries that use stable, widely understood technology to make relatively simple products or high-technology industries that use rapidly developing technology to make a continuous stream of quickly obsolete new products.

The traditional industries have generally been associated with fairly labor-intensive technologies. Because of persistently low wages in the Third World and the relatively low investment needed to begin to produce these products, these industries have led the burgeoning exports of manufactures produced in developing countries.

The high-tech industries, in contrast, depend for their success on access to the specialized resources required for research and development and highly complex production processes. These industries have therefore been located in the industrialized countries. As products mature and technology diffuses, high-tech products eventually become traditional products, and production moves to more competitive locations abroad.

The phenomenon of production abroad may be viewed as a system geared to retaining competitiveness for firms in developed countries after a product has entered the downside of the product cycle. That is, the firms that developed the product continue to produce profitably by eventually relocating or subcontracting assembly production facilities in low-wage developing countries.

When this strategy works, these firms generally have some other competitive cost advantages, such as access to capital, marketing administration, or technology, since an indigenous firm producing standard products in its native business environment presumably could do so at no greater and possibly at lower cost. Also, production processes must permit such a division of labor, and transportation costs should not be an excessive component of total costs.

Growth in Production Sharing

Assembly abroad has grown dramatically during the last few decades. Before wage differentials became an important factor in world trade, coproduction, linked to technology and skill specialization, was primarily a phenomenon of industrialized countries. Thus, production sharing among industrialized countries has involved sophisticated goods with technologically advanced production processes. Production sharing between industrial and developing countries is a comparatively recent phenomenon, and has been stimulated by improvements in transport and communications.

The emergence of overseas assembly activities was a natural result of growing worldwide competition in the manufacturing industry in the postwar period. As Western Europe recovered and Japan quickly became an important industrial power, the United States was the first to face new competition because its wages were so high relative to those throughout the rest of the world. As domestic labor-intensive production became less and less economical, U.S. firms began breaking production into stages and carrying out the labor-intensive processes in countries where wages were low. In response to essentially the same conditions, the industrially advanced countries of Europe imported low-cost laborers, and Japan turned to automation when wages rose.

Although firms in all industrialized countries have engaged in foreign assembly, U.S. firms have done by far the most. Because of a higher degree of protectionism, European countries have carried out comparatively little production abroad. In Japan, such activities have been used primarily as a means of penetrating foreign markets, although it appears that more recently Japanese companies have reimported a growing volume of products assembled for them abroad. On the other side of the coproduction relationship, the principal participants are developing countries in the Far East and the Caribbean Basin.

Tariff Provisions 806 and 807

Products assembled abroad reach the U.S. market under tariff items 806.30 and 807.00 (henceforth 806/807), which permit the duty-free entry of U.S. components sent abroad for processing or assembly.* Since 1966, U.S. 806/807 imports, which totaled $22 billion in 1983, have grown faster than the rapidly growing U.S. import bill, rising from less than 4 percent to almost 10 percent of total U.S. imports of merchandise in 1983. The growth has been much more rapid in relation to imports of selected groups of manufactures. U.S. imports under sections 806/807 are largely confined to motor vehicles and parts, apparel, and various types of electrical equipment. Almost all the motor vehicle imports come from industrialized countries, and almost all other 806/807 imports come from developing countries.

Mexico's share of U.S. 806/807 imports has risen dramatically since 1969

*This section is drawn from Flamm and Grunwald (1985).

TABLE 1 U.S 806/807 Imports from 15 Countries, 1969 and 1983 (millions of dollars)

	1969		1983	
Country of Assembly	Total 806/807 Imports	Value of Duty-Free U.S. Components	Total 806/807 Imports	Value of Duty-Free U.S. Components
Industrial countries				
Federal Republic of Germany	627.4	11.6	2,736.7	58.9
Japan	137.9	25.4	6,489.6	175.5
Canada	340.1	118.7	1,425.9	467.0
Developing countries				
Mexico	150.0	97.9	3,716.9	1,908.7
Malaysia	0.4	0.1	1,203.2	695.7
Singapore	11.6	3.8	983.3	275.9
Philippines	5.2	3.5	725.9	455.6
Republic of Korea	23.8	15.9	575.6	340.4
Taiwan	68.7	23.8	568.5	100.7
Hong Kong	91.4	51.3	448.1	72.2
Haiti	4.0	2.4	197.4	139.4
Brazil	4.1	2.5	193.1	27.4
Dominican Republic	0.1	0.1	161.0	111.6
El Salvador	0.2	0.1	79.6	45.3
Colombia	0.4	0.2	29.8	20.0
Fifteen countries listed above	1,465.3	357.3	19,534.6	4,894.3
All countries	1,841.8	442.6	21,845.7	5,447.1

SOURCE: U.S. Tariff Commission (1970, pp. A-57, A-87), and special tabulation from the U.S. International Trade Commission, 1983.

(see Table 1). In that year Mexico's share was 8.2 percent. By 1983 it reached 17 percent, surpassing almost three to one the imports from any other developing country. Mexico is now second only to Japan in its share of total U.S. 806/807 imports.

Organizational Arrangement of Production Sharing

Much of the 806/807 trade is internal to the multinational operations of U.S. firms. In 1969 the U.S. Tariff Commission found that more than half the value of duty-free components reimported under 806/807 tariffs came from U.S.-owned investments. The remainder came from U.S. firms dealing with independent contractors and jobbers abroad and from foreign firms securing U.S. components for their exports to the United States. In most respects, this trend has continued into the 1980s. The principal exception is the apparel industry. In 1969, according to the U.S. Tariff Commission, about four-fifths of the apparel imported under 807 was assembled by contractors or jobbers dealing with foreign subcontractors in which they had no financial interest.

EVOLUTION OF PRODUCTION SHARING FROM THE POINT OF VIEW OF THE DEVELOPING COUNTRY: THE MEXICAN CASE

Factors Contributing to Mexico's Importance as a Production-Sharing Site

It is not surprising that Mexico has become an important partner in assembly activities abroad, since U.S. firms can gain substantial competitive advantages by operating there. At least three major factors led to this situation.

Proximity is one such factor. Mexico, a developing country, shares nearly 2,000 miles (3,200 kilometers) of border with the United States, a country that has one of the highest wage rates and is by far the largest producer in the world. The border is easily accessible, and transportation to any other overseas trading partner is available. In addition to the proximity of the two countries, they share similarities in geography and culture. Many U.S. entrepreneurs and business executives have conducted business in Mexico, and for historical reasons as well as through migration, Mexico's culture is diffused throughout many parts of the United States. Many Mexican professionals and government officials are fully bilingual, so business in Mexico can be carried out almost completely in English.

The main advantage derived from the location of new assembly operations, called *maquiladoras,* in Mexico is low-cost access to the U.S. market. A site in Ciudad Juarez is about the same distance from U.S. markets as a Chicago site is (see Figure 1). Similarly, a site in Tijuana is hard to improve upon as a point from which to reach the southern California market.

Shorter distances translate into still other advantages. Because products can take as little as 2 or 3 days to reach their destination, inventory cycles can be cut considerably. Short travel distances also provide operational advantages, the most important of which are low investment in warehouse space and greater production flexibility. Also, because Mexico and the United States are in the same hemisphere, production-related communications that have to be handled indirectly when working with plants located in the Pacific Rim sites can be dealt with by telephone calls between managers in U.S. plants and producers in Mexico.

Demography is the second factor that has made Mexico an important production-sharing partner. The United States, like all developed countries, will experience a shortage of young people entering the labor force and available for traditional manufacturing jobs. A few figures illustrate this

MILEAGE

FROM CHICAGO TO:		FROM CIUDAD JUAREZ TO:	
NEW YORK	831	NEW YORK	2,299
LOS ANGELES	2,113	LOS ANGELES	790
DALLAS	933	DALLAS	625
SAN FRANCISCO	2,205	SAN FRANCISCO	1,170
ATLANTA	722	ATLANTA	1,694
AVERAGE	1,351	AVERAGE	1,316

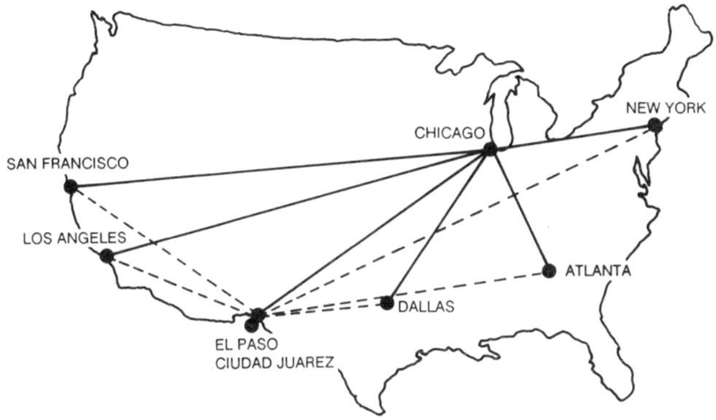

FIGURE 1 Mexican transportation advantage.

trend. As of 1984, the U.S. economy employed 105 million people. Employment since 1975 has been growing approximately 2.2 percent annually. For this period, the average number of jobs created in the American economy has been approximately 2.1 million per year. However, the population of 18- to 65-year-olds will increase at a rate of only 0.8 percent per year for the rest of the century, with 1.2 million people annually reaching working age. Thus, on average, if the patterns of the past 10 years prevail, the number of jobs created will exceed the number of people entering the work force by approximately 900,000.

In sharp contrast to the situation in the United States, about 20 million people were employed in Mexico in 1986. However, the number of 18- to 65-year-old people will increase by nearly 3 percent per year for the rest of the century, with an average of 1.5 million people annually entering this age group. Thus, each year, 300,000 more people will reach working age in Mexico than in the United States.

The U.S. economy is currently about 20 times the size of the Mexican economy (see Figure 2). However, during the rest of the century, the Mexican

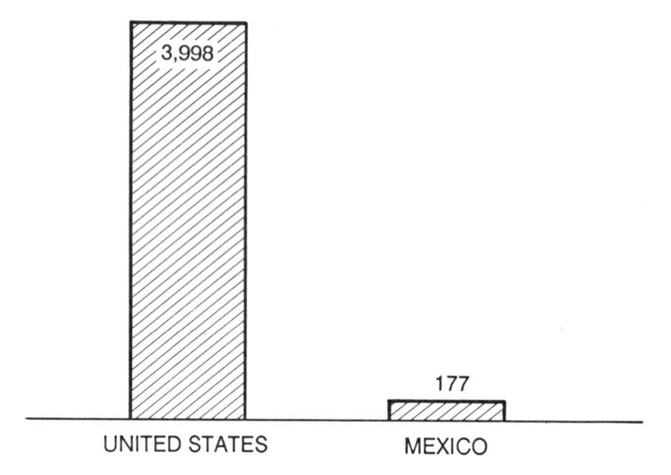

FIGURE 2 Relative size of Mexican and U.S. economies, 1985 (billions of dollars).

economy will have to incorporate into its labor force more people than the U.S. economy. Thus, the most pressing political problem for Mexico will be the creation of jobs. To avoid extensive migration from Mexico and even the possibility of unrest, U.S. policymakers should also favor job creation in Mexico. This imbalance between the two countries in numbers of jobs and workers available will create a large wage differential that will persist for many years to come.

The present Mexican economic crisis is another factor that makes production sharing an attractive option. Mexico's economic development strategy during the 1950s, 1960s, and early 1970s was characterized by a stable, annual 6 percent gross national product (GNP) growth rate without significant inflation. This was achieved by a public spending policy that kept government finances in equilibrium by assigning the private sector an enhanced role in economic activities. Equally important was that a trade policy based on import substitution permitted the creation of a major industrial infrastructure throughout the country and also financed the education of the majority of the Mexican population. However, the Mexican government had to change this development strategy in the 1970s because it did not result in the transfer of greater numbers of people from redundant, informal economic activities, such as street vending, into productive jobs.

In an all-out effort to promote job creation in both the public and the private sectors, the governments of Luis Echeverria (1970–1976) and José Lopéz Portillo (1976–1982) increased the level of public-sector spending through greater budget deficits and foreign debt. However, the results achieved were minor. Despite average GNP growth slightly higher than that in the previous development period, the Mexican economy started on increasingly complex cycles of stop-and-go growth and dangerous levels of inflation.

The collapse of oil prices brought the present foreign debt crisis and the need to restructure the economic development strategy of the country. After the crisis exploded in 1982, the government implemented a stabilization program that included measures to increase the attractiveness of production sharing. The first was a general trend toward opening the economy and eliminating trade barriers. This trend has culminated in Mexico's joining the General Agreement on Tariffs and Trade (GATT) and gradually eliminating quotas and import permits for duties in the import sector.

Second, the peso was devalued severely. Between mid-1981 and December 1982 the peso value fell from four U.S. cents to sixth-tenths of a U.S. cent, an 83 percent devaluation. Consequently, during this short period, the peso went from being seriously overvalued (by 30 to 40 percent) to being seriously undervalued. It has stayed undervalued ever since, through continuous slippage of the parity rate of the peso. Consequently, the real cost of unskilled labor has declined and the cost of labor has become competitive with labor costs in other newly industrializing countries (see Figure 3).

Simultaneously, the economy entered a drastic recession that depressed the cost of skilled labor of all types, including professionals and managers. Thus, additional advantages can be seized by hiring highly qualified workers, including managers and technicians, at a fraction of the cost of those in the United States or elsewhere.

As a result of the crisis, other key U.S. imports have also become extremely

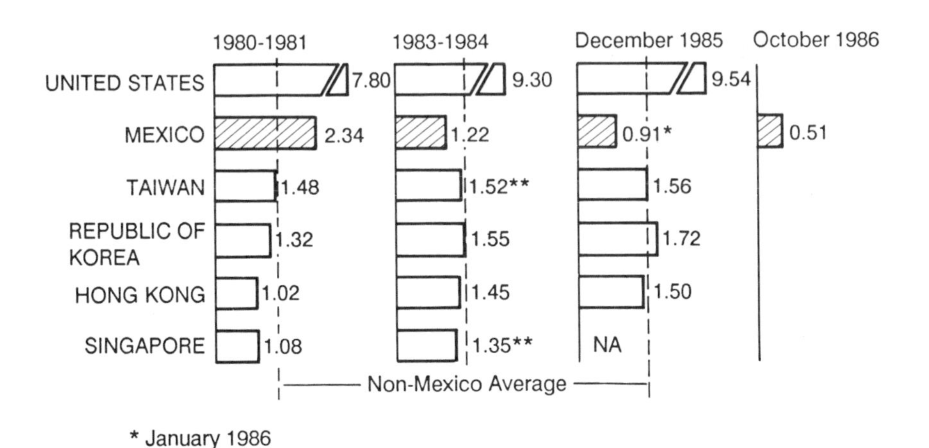

FIGURE 3 Mexican industrial hourly wage rate advantage, including benefits (dollars per hour, unskilled labor). NA indicates not applicable. SOURCE: International Labor Organization and International Monetary Fund.

competitive. Energy is abundant and inexpensive in Mexico, as are many other industrial inputs that Mexico supplies worldwide. Examples of these are copper, silver, lead, zinc, ammonia, polyvinyl chloride, ethylene, fertilizers, cement, glass, cotton, essential oils, and coffee.

Organizational Arrangement for Production Sharing

The Mexican authorities realized the potential for production sharing in the mid-1960s. At that time, Mexico instituted its border industrialization program to increase the advantage of its proximity to the United States. This program was aimed at absorbing the potential border unemployment arising from the 1964 termination of the *bracero* or temporary worker program between the United States and Mexico. It allowed duty-free entry of machinery and components for processing or assembly within a 12-mile (20-kilometer) strip along the border, provided that all imported products were reexported. Thus, none of the output of assembly operations could be sold in Mexico.

Subsequent Mexican legislation and administrative regulation expanded the scope of the *maquila,* first by exempting the *maquiladoras* from the Mexicanization requirements of Mexican majority ownership, and second, by permitting the establishment of *maquiladoras* anywhere in the country, subject to approval by the authorities. Foreign technicians and managerial personnel may reside in Mexico if their presence is considered necessary for the efficient functioning of the *maquiladoras*. Finally, as a part of the *maquiladora* promotion decree of August 1983, a firm may sell up to 20 percent of its production in the Mexican market if it complies with a minimum Mexican content requirement, established on a case-by-case basis by the Mexican Secretariat of Commerce and Industrial Development. Customs procedures and other government formalities have been eased, as a further attraction to *maquila* operation. Industrial parks have been promoted along the border, as well as in the interior.

Production organization has taken three main forms. The first involves subcontracting with another Mexican firm to assemble an item to specification. With this arrangement, both investment requirements and institutional barriers are low, and the risk of the operation is therefore kept to a minimum.

The second configuration is the so-called shelter operation. This is actually a joint venture in which the shelter provider puts up real estate, installs the plant, and provides day-to-day management, whereas the foreign firm provides equipment, specialized installations, and crucial know-how. This arrangement best uses the complementary skills of each party and consequently may be the strongest long-term configuration.

Maquiladoras fully owned by foreign parent companies, however, remain the most prevalent form of production sharing in Mexico today and account for the majority of the *maquila* output. Since 90 percent or more of the output is for the U.S. market, most of these subsidiaries are U.S. owned or controlled.

In the past, most of the U.S.-controlled *maquiladoras* were subsidiaries of medium-size multinational enterprises. More recently, however, some of the giants in U.S. industry have established assembly operations in Mexico. At least 100 of the Fortune 500 companies had *maquiladoras* as of 1986. In addition, three of the largest Japanese firms and four of the top European consortia operated assembly plants in Mexico as long ago as 1978.

Growth of Production-Sharing Operations in Mexico: Size, Product Mix, and Location

Since 1965, when the border industrialization program got under way, a significant number of assembly plants have been established almost every year. In 1986 there were 844 plants, employing 250,000 Mexicans. Although this number is significant, the *maquila* labor force is only slightly more than 1 percent of total Mexican employment.

The value added by the *maquiladoras* in Mexico has been increasing sharply since the beginning of the border industrialization program. In 1986 it reached $1.285 billion, surpassing tourism and border transactions as a source of foreign exchange for Mexico.

Mexican assembly activities cover increasingly wide ranges: from toys and dolls to sophisticated electronic equipment (see Figure 4). Most of the plants are concentrated in six towns along the border, from Tijuana on the Pacific Ocean to Matamoros opposite Brownsville, Texas, near the Gulf of Mexico. Plants concentrate along the border to minimize the distance between assembly plants and their supply sources and markets, which are in the United States. Border towns also have a very bilingual population, and the labor force has a large cultural influence from the United States, which makes it easy to train workers in modern methods. In 1983 about 11 percent of the total, or 67 plants, were in the interior of the country. This proportion, although small, has been increasing because job creation at the border has created labor shortages in these towns and wages are increasing faster there than they are in the interior of the country. As good communication and transportation facilities become available in the interior, firms will find the labor cost differential attractive and will increasingly locate in the interior of the country. A case in point is Yucatán, which by air or sea is close to the East Coast of the United States and is experiencing a healthy growth of *maquila* activities. The plants are now situated throughout the country, including the three largest cities of Mexico City, Guadalajara, and Monterrey.

Issues of Assembly Operations in Mexico

Because so many foreign firms are involved in Mexican assembly activities, Mexicans and others have extensively debated its merits for the country.

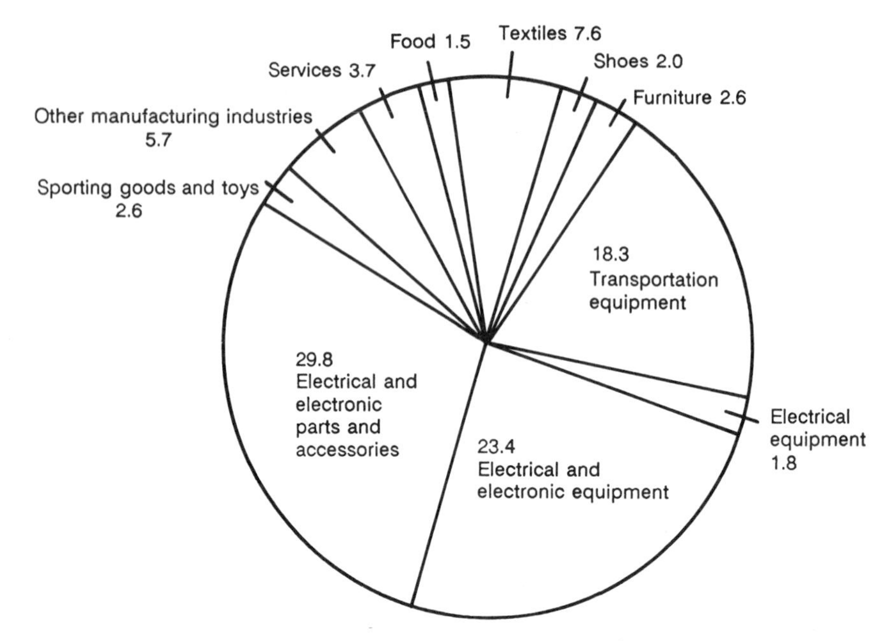

FIGURE 4 In-bond industry value added by activity, 1982 (percentage of total value added). Classification of activity. SOURCE: Mexican Ministry of Programs and Planning.

Benefits have been questioned, and serious negative effects on the Mexican economy and society as a whole have been attributed to the assembly arrangements. The critique centers on three principal issues:

- the absence of significant linkages of assembly activities to the Mexican economy;
- the effects on the labor force and on society in the areas where *maquiladoras* are concentrated; and
- the vulnerability of *maquiladoras* to swings in the U.S. business cycle and their general dependence on decisions made outside Mexico.

Linkages with the National Economy A persuasive argument can be made that the foreign-owned assembly services have not extended their benefits sufficiently to the rest of the Mexican economy. First, only a trivial percentage of the materials used in these operations is of Mexican origin. That percentage has hovered around 1.5 percent of the total use of components and supplies during the period from 1975 to 1983. In the plants in the interior, however, the use of domestic materials has been considerably greater, ranging from 4 to 15 percent of total materials used. Next, most of the jobs created are for unskilled workers. The labor force, therefore, receives little training. Although many of the assembly plants use sophisticated equipment and technology, there is a low level of technology transfer to the rest of the Mexican economy. Furthermore, only part of the wages paid to assembly workers are spent on Mexican goods and services, because the population near the border routinely shops across the frontier for a significant proportion of its requirements. Thus, it is argued that the income generated by assembly production for foreign manufacturers will provide only a limited stimulus to the Mexican economy.

There is little doubt that linkages can be improved. The experience of Hong Kong, the Republic of Korea, Taiwan, and Singapore seems to prove that assembly activities will eventually integrate assembly production into the national economy. However, to break the vicious cycles that have prevented further linkages of the *maquiladora* operations to the rest of the economy, specific policy actions must be undertaken. This issue will be discussed later.

Employment According to several recent studies, the employment generated by the *maquiladoras* does not absorb traditional unemployment and underemployment. A striking aspect of the labor force in Mexican assembly industry is that women represent more than three-quarters of the total (see Figure 5). Most are quite young and have not previously been in the labor force. Relatively few of the female *maquila* workers are heads of households and most provide only supplementary income for their families. Thus, *maquila* workers in Mexico are drawn from a sector of the population that has never worked or sought work before. Therefore, one of the main original objectives of the Mexican border industrialization program, which was to absorb the rural male migrant workers left stranded by the termination of the

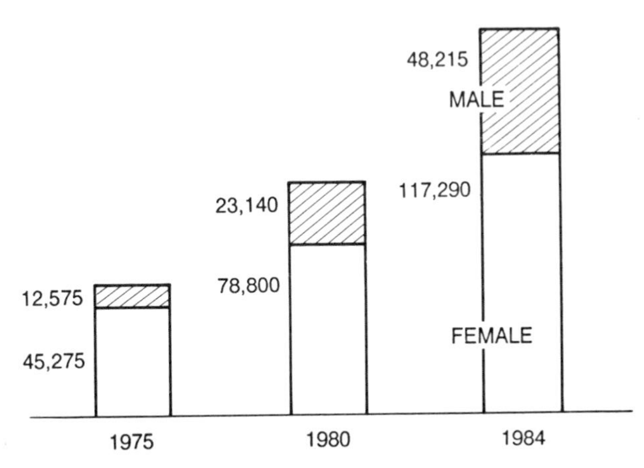

FIGURE 5 Number (in thousands) of employees, by sex, in *maquila* activities. SOURCE: Mexican Ministry of Programs and Planning.

bracero arrangement with the United States, has not been accomplished by the establishment of assembly operations. The preponderance of female workers in assembly production, however, is not confined to developing countries. A greater proportion of women in the labor force is a universal phenomenon of modern economic development. Social adjustments of this type are painful, but they are a part of the contemporary process of economic development. The introduction of assembly activities into particular regions of Mexico has accelerated the need for this kind of adjustment.

The Dependency Issue The Mexican assembly industry was affected to some degree by the U.S. recession of 1974–1975. Some *maquiladoras* closed down, and many workers were laid off. The effect of the recession was concentrated in the *maquiladora* operations of the electrical and electronic assembly industries. Apparel and textiles industries were not affected; employment in these sectors even increased slightly in 1974–1975. Although the effect of the recession was swift and severe, assembly activities bounced back relatively quickly.

A sharp rise in Mexican wages aggravated the effects of the U.S. recession during that period. Mexican wages in dollar terms increased because the parity between the peso and the dollar was kept constant, despite a large inflation differential between Mexico and the United States. Wage increases continued to outpace Mexican inflation until 1976, reaching a level 24 percent above the 1970 level in real terms. Wage costs in assembly firms rose with the nominal wage increases. Minimum wages in U.S. dollars doubled between 1971 and 1976. It is therefore not surprising that many of the less efficient assembly operations could not bear the cost pressure and had to close. However, after the devaluation of the peso in 1976, employment levels quickly recovered.

Although the growth of the assembly industry in Mexico after 1982 can be related in part to the devaluation in that year, the effects of the recent recession in the United States confirm the conclusion drawn from the experience of the mid-1970s: The downward trend of U.S. business cycles is reflected to a much smaller degree, if at all, in Mexican assembly activities for U.S. firms. The shakeout of unstable firms took place, but output and employment expanded with fewer plants, as U.S. companies transferred more of their labor-intensive operations to low-cost locations abroad.

In summary, the dependency issue merits concern. Assembly production may be more sensitive to external economic conditions and decisions than are some other economic activities within Mexico. After all, foreign assembly has been directed exclusively toward exports, and it is natural that sales will depend on foreign markets. These activities are also tied to foreign decisions, either through the operations of foreign subsidiaries or through international subcontracting. However, declines in assembly operations that were attributed to U.S. recessions were relatively mild and short-lived. These effects were particularly small compared to the decline in Mexican manufactured exports during this period, not including assembly, which dropped more than 14 percent in 1975.

Policy Implications for the Developing Countries: The Case of Mexico

Assembly production is extremely significant for developing countries. Although it is not the primary engine of economic development, the assembly sector has been an important generator of income, employment, and foreign exchange. Despite their positive and important contributions, however, assembly activities have been regarded with some ambivalence by policymakers of developing countries. Although it is clear that national enterprise in Mexico has not gained as much as it could from assembly activities, it is unlikely that Mexico would be better off without them. The alternatives to employment in assembly activities—unemployment or underemployment—are generally much less desirable. Given the availability of large, underused resources, assembly production offers the host country significant economic opportunities.

There should be two main policy objectives regarding production sharing. One is to preserve and enhance the competitive advantages of Mexico as a production-sharing site; the other should be to promote a tighter linkage of assembly activities to the rest of the economy.

To enhance the competitive advantage of Mexico, it will be necessary to continue implementing the following policy guidelines:

- Competitive exchange rates and salary structures. Growing evidence suggests that assembly activities are highly competitive in the global environment and that the international outsourcing of production by firms is responsive to cost changes. Therefore, if Mexico wants to promote growth in its *maquiladora* sector, it will have to continue pursuing competitive exchange rate policies.
- The streamlining of administrative procedures. This deserves a high priority. Bureaucratic rigidities in the issuance of permits by regulatory authorities have fostered delays in both establishing and operating firms and should be tenaciously combated.
- Enhancing its transportation and telecommunication infrastructure. This will permit fuller exploitation of the advantages of proximity and culture that Mexico has with its key production-sharing market. Multinational firms will find the facilities to which they have become accustomed in their domestic dealings. In the case of Telefonos de Mexico, a special effort has been made to provide state-of-the-art telecommunications to

all the industrial parks with significant assembly activities, offering quality levels and service similar to those across the border.

In linking the *maquiladora* industry to the rest of the Mexican economy, it can be affirmed that existing policies of free trade zones, in-bond arrangements, waiving of requirements of national ownership and content, and similar measures are designed to expand assembly operations. However, further specific policies should be pursued to better integrate the *maquila* sector to the rest of the country. Support of assembly activities should be aligned with the general trade policy of the country. Mexico is dismantling a protectionist trade policy that has been in place for almost 40 years because its present macroeconomic condition foresees a continued shortage of foreign exchange, and also in light of the faster growth experienced in newly industrialized countries that have opted for an open economy. Mexico has recently entered the GATT and is actively substituting quotas and import permits for duties. Mexico is also promoting direct foreign investment and embarking on an all-out effort to develop its non-oil export sector.

These policies will affect assembly production very favorably. Heightened competition arising from trade liberalization will force national firms to reduce costs and increase quality, and therefore facilitate greater participation of national enterprises in assembly activities. With lower costs, firms could reduce prices of products that could serve as inputs for the assembly plants. Competition would also raise their standards of performance so they would be able to meet the more stringent requirements of export production, as both potential suppliers to and operators of assembly plants. Thus, trade liberalization policy might help increase the percentage of Mexican materials in assembly products and the participation of local capital in export activities.

Along with trade liberalization, special incentives might be needed to induce national capital to venture into assembly activities as suppliers or operators. A strong inducement would be to eliminate all restrictions both on production for the domestic market and on assembly for exports. This would permit the fuller use of capacity in existing plants and reduce the risk to national capital of investment in assembly plants. Fluctuations in both domestic and foreign business can be smoothed if operations can be shifted between production for the domestic market and production for export.

The attractiveness of assembly activities to local enterprises could be enhanced further by lifting the restriction on the sale of in-bond assembly products to the domestic market. Although these sales would not be exempt from customs duties on imported materials, they would make assembled items available to the domestic economy at a lower cost than if they had to be reimported after their return to the United States. Permitting the use of in-bond subassemblies as inputs for goods destined for the domestic market could also benefit producers as well as consumers by lowering costs and increasing the quality of the final goods.

One step in the right direction is the recent regulation that permits local sales of up to 20 percent of total production in exchange for an increase in the national content of the product. Firms willing to increase local content will be rewarded with domestic sales as they work to make local enterprises more efficient suppliers of components to assembly plants. Indigenous industry, once it becomes a supplier of assembly plants, will support greater access to local markets by foreign assembly subsidiaries. At the same time, the relaxation of the export requirements of in-bond production will attract national capital into assembly production.

Issues of Production Sharing in Developed Countries

The transfer of labor-intensive operations abroad benefits the industrial countries by lowering the cost of assembled goods to the consumer. However, it alters the wage and profit rates in the domestic economy and fosters the equalization of factor prices internationally.

Import competition reduces the incomes of specific workers in exchange for general improvement in economic well-being. The obvious objective of U.S. firms in transferring assembly work abroad is to reduce the cost of production, which means lower prices to consumers. If the estimated cost savings from Mexican operations amount to approximately $10,000 per worker, the total direct savings to the U.S. economy from Mexican assembly production was roughly $2.5 billion in 1986. Although this represents less than 1 percent of U.S. value added in manufacturing, savings in cost on individual products are substantial, and the gain in welfare to purchasers, although concentrated in only a few products, is significant. For producers, small differences in cost can spell the survival of a firm. There is evidence that production arrangements with other countries have allowed certain businesses to remain internationally competitive in the face of serious threats from imports.

In exchange for these benefits to an industrial country, the displacement of workers in an industry going abroad is the principal negative effect. Unskilled operators in manufacturing jobs made up only about 8 percent of U.S. employment in 1978. The continued displacement of these unskilled workers will mean the end of their relatively high income in Western industrialized nations. Trade theory predicts that wages as well as other factor prices will be equalized across countries, and continued downward pressure on real wages of the unskilled manufacturing workers seems likely.

Policy Implications for the Developed Countries

Recent and growing criticism of free trade in general, and offshore production sharing in particular, centers around a strong ideological belief that the key objective of trade policy is the maintenance and creation of jobs.

Adopting only this perspective, a more protectionist trade policy could perhaps be justified for the short term. However, for the wealthiest economy in the world, there are other key objectives that should also be pursued by trade policy and that clearly favor the increase of world trade and production sharing. Among these are:

- an increase in the welfare of society at large through access to cheaper and better products and services through foreign trade; and
- a geopolitical advantage that the United States and the Western world derive from advancing peaceful development.

Protectionism, in its search for job protection, deprives the majority of society of the cheaper goods that free trade brings and will also create insurmountable tension among countries. Also, protectionism may deprive the less developed countries of the foreign exchange they need to import essential goods to their economies and service their debts. This will result in further economic stagnation and greater political instability, with far-reaching effects for the welfare of Western society. Possible consequences include a threat to the world financial system. More countries will be forced to act unilaterally and turn their backs on the Western economic system in search of internal political support. They will then be forced to search for support in other geopolitical forces, including nondemocratic political regimes. Finally, greater foreign migration will be unavoidable, regardless of restrictive legislation and massive policing of borders.

Inevitably, market pressures will cause a fall in prices of abundant resources. The growth of the world's labor force will increase, according to some estimates, by 50 percent in the remaining years of this century. An additional 750 million people will enter the labor force by the year 2000, of whom 680 million will come from Third World nations. Developed economies must continue upgrading their labor forces to avoid competition that would erode the standard of living of their less-skilled laborers. Therefore, they should institute a system of transfers to ease the pain of displaced workers who are forced to move, search for new employment, and acquire new skills, so that they will not have to bear the total weight of socially desirable changes in trade and manufacturing arrangements.

At the same time, developed countries should keep striving to maintain a balanced trade pattern by keeping international exchange rates flexible and correcting huge trade deficits. Also, an effort should be made to reduce the economic dependence of debtor nations. Efforts of debtor nations to earn dollars to service their debt have caused American exports to these nations to decline precipitously. These debtor countries that must maintain their balance of payments in equilibrium, in spite of having to serve the tremendous burden of their foreign debts, are forced to export more goods and import fewer goods. A long-term adjustment of debt-service requirements would enable debtor nations to import more from the United States and other developed economies and simultaneously to focus development efforts on reinvigorating their internal markets and strengthening domestic forces for long-term economic growth.

It is also in the best interests of developed economies to institute policies that promote the reorganization of industry and the redeployment of production stages in which developed countries are no longer competitive. If an orderly process of production reorganization is deemed desirable, developed countries should give high priority to removing barriers such as quotas and voluntary restraints that are particularly burdensome to the poor countries.

U.S. policy should also promote the streamlining of tariffs 806/807 to make those provisions more efficient and better link assembly activities to the domestic host economy, without diminishing the incentives to use U.S. materials. Tariff 807 could be broadened to include operations such as cutting, mixing, and processing of U.S. components abroad, while still permitting them to be brought back to the United States duty free. This would also extend the opportunities for local participation in assembly production, and at the same time, it might also broaden the use of U.S. components.

An additional step to the creation of linkages would be a selective application of a removal of duties from goods assembled in countries where the United States has key geopolitical interests. Thus, including the tariff exception, the value added abroad would encourage greater use of foreign inputs. This would ultimately benefit U.S. consumers with lower-priced goods.

Developed countries should also fight to dispel the image of assembly production as an enclave in the host economy, an image borne out by the strikingly small percentages of local materials used in these operations. A special effort should be made to provide incentives for the United States and developed country subsidiaries to provide the necessary technological expertise that domestic suppliers need to efficiently supply assembly operations.

It is also clear that subcontracting with local firms in developing countries is more conducive to technology transfer and the establishment of linkages than is operating assembly plants abroad through U.S. subsidiaries. The subcontractor in the developing country not only benefits directly from working with the U.S. principal but may also transfer the knowledge acquired to production for domestic markets, thus spreading the effects.

CONCLUSIONS

The world economy has passed successively through more comprehensive stages of integration. Markets for traded goods, and more recently, financial markets, have merged across national boundaries into a truly international structure. Labor markets are also in the process of being integrated into an international system, partly through migration, but mainly through trade in the changing global pattern of the international division of labor. Coproduction can be seen as a principal feature of the international organization of industry. Firms in industrially advanced countries collaborate with firms in developing countries to manufacture a variety of products to the mutual benefit of both countries.

The implication of much of what has been described here is that policy should smooth the way for more sophisticated coproduction stages between industrial and developing countries. As developing countries advance, they should be able to increase the levels of subcontracting, local capital participation, and national materials in assembly activities. As long as sharp differences persist between wages in economically advanced countries and those in less developed regions, the rationale for assembly production abroad will remain. The significance of the international reorganization of product flows within a single industry is that the present high wages for unskilled labor in the United States and other industrialized countries will no longer be insulated from international competition. Technological advances in the United States are unlikely to offset this trend, so that what remains is a fundamental long-term policy choice for industrial societies. One choice is to face a reduction in the comparatively high standards of living of the least skilled, and with it, a period of social and political turbulence. The other alternative is to deliberately eliminate those sectors of the labor force that are in direct competition with workers in developing countries through investment in education and training. This, in effect, shifts the composition of the labor force to more highly skilled occupations. Given the social cost of an unplanned transition that suddenly eliminates the unskilled workers from the industrial work force, large-scale investment in upgrading the U.S. work force should be the preferred choice.

From the Mexican point of view, a coherent medium-term development policy should be aimed at creating jobs for the rapidly growing labor force. Job creation should be the number-one priority of Mexican policymakers. The *maquiladoras* and production sharing should be one of the main pillars of this strategy. Not only are they labor intensive and big foreign exchange earners, but the capital investment needed to create a job in this sector has been estimated to be around six times lower than that in the traditional manufacturing sector. Therefore, both countries should adopt production sharing enthusiastically as an alternative to be fostered instead of a malady to be remedied.

BIBLIOGRAPHY

American Chamber of Commerce of Mexico, A.C. 1986. Mexico's *Maquiladora* In-Bond Industry Handbook. México, D.F.

Arthur D. Little Mexicana, S.A. de C.V. 1986. Crisis y Apertura Internacional: Oportunidades y Amenazas para el Empresario Méxicano. México, D.F.

Drucker, P. 1969. The Age of Discontinuity: Guidelines to Our Changing Society. New York: Harper & Row.

Drucker, P. 1980. Managing in Turbulent Times. London: Pan Books.

Drucker, P. 1986. El Cambio en la Economía Mundial. Mercado de Valores 25 August 1986, 811–816.

Flamm, K., and J. Grunwald. 1985. The Global Factory: Foreign Assembly in International Trade. Washington, D.C.: Brookings Institution.

McKinsey and Company, Inc. 1987. Shaping effective responses to the globalization challenge. Unpublished manuscript.

Ohmae, Kenichi. 1985. Triad Power: The Coming Shape of Global Competition. New York: The Free Press.

Tellez, L. 1986. Essays on Real and Financial Aspects of an Open Economy: The Mexican Case. Ph.D. dissertation, Massachusetts Institute of Technology.

Telmex. 1986. Telecommunications Services for the *Maquiladora* Industry: A TELMEX Future Perspective. México, D.F.

U.S. Tariff Commission. 1970. Economic Factors Affecting the Use of Items 807.00 and 806.30 of the Tariff Schedules of the United States. Publication 339. Washington, D.C.: U.S. Tariff Commission.

Regional and National Consequences of Globalizing Industries of the Pacific Rim

JAN E. KOLM

Globalization of industries—the spread of industries throughout the world—is as old as history. Examples abound: the transfer of glass-making from Syria to Italy, the Rhine, and Bohemia; of metallurgy from Rome to Gaul, Spain, and Cornwall; of silk from China to Turkey, Italy, Central Europe, and England.

Equally old are the forces opposing it—inertia and protectionism. To cite one example, the Venetians were past masters of trade, not only in rare wares—spice and treasures from the East (high-added-value goods)—but also in the import of technologies, such as the making of fine glasses, silk robes, and laces. Having acquired these skills, they were eager to keep them exclusive, as is evident from their Senate's edict:

> If any artist or handicraftsman practices his art in any foreign land to the detriment of the Republic, orders to return will be sent to him; if he disobeys them, his next of kin will be put to prison. . . . If he comes back, his past offences will be condoned and employment will be found for him in Venice, but if notwithstanding . . . he obstinately decides to continue living abroad, an emissary will be commissioned to kill him and his next of kin will be liberated upon his death. (Earnshaw, 1980, p. 8)

What has changed in globalization is the magnitude and pace of activity. The settled world is larger; rates of growth and technological change are faster; and people, information, and capital are all more mobile.

Both the transfer of developed technologies into new regions and the protection of infant industries by tariffs or subsidies are mechanisms of growth. Indeed, the key source of new technologies, research and development (R&D), can be viewed as protectionism. To reach a minimum viable scale, R&D must be subsidized from the cash flow of the mature economic unit, which may be the state or, in a large transnational enterprise, the mature business.

THE PACIFIC RIM: A THEORETICAL CONSTRUCT

The Pacific Economic Community (PEC)[1] (excluding Latin America and the People's Republic of China) extends over one-fifth of the earth, embraces nearly one-sixth of its population, and is characterized by a wide diversity in population density; natural endowments; state of development; race, religion, language, script, and culture; and isolation from and proximity to world population centers. Although not a formally constituted community like the European Economic Community (EEC), the PEC is the world's most rapidly growing trading area. It includes several states that developed unusually quickly, shifting from undeveloped to industrialized nations in a few decades. Although the region remains heterogeneous, internal bilateral and multilateral linkages are gradually developing. Also, surprisingly in a region with a predominantly Asian population, English is becoming its *lingua franca*, and so is providing an additional cultural link.

In this necessarily simplistic overview, the dynamics of economic development in this diverse group of countries can perhaps best be interpreted in terms of the technological complexity of goods and the product cycle, that is, the cycle of shifting comparative advantage (Vernon, 1966).

Goods and countries can then be classified as follows:

Goods	*Countries*
• Raw materials—natural resource-intensive goods, very low technology content	• Undeveloped countries (UCs)
• Refined goods—labor-intensive technologies, low technology content	• Developing countries (DCs)
• Manufactured goods—capital-intensive technologies, high technology content	• Newly industrializing countries (NICs)
• Processed and capital goods—postindustrial high technologies and services, very high technology content	• Industralized countries (ICs)

The Product Cycle

The product cycle begins in the industrialized, innovating country. The technology is diffused through exports, foreign investment, and licensing. This produces a cascading flow from highly developed to less developed countries, with progressive loss of comparative advantage to the low-labor-cost countries. The impact on the export-import balance between countries at different stages of development can be presented schematically as shown in Figure 1.[2]

Superimposed on the product cycle is obsolescence, the progressive decay of products. With increased volume and ease of production, this decay leads

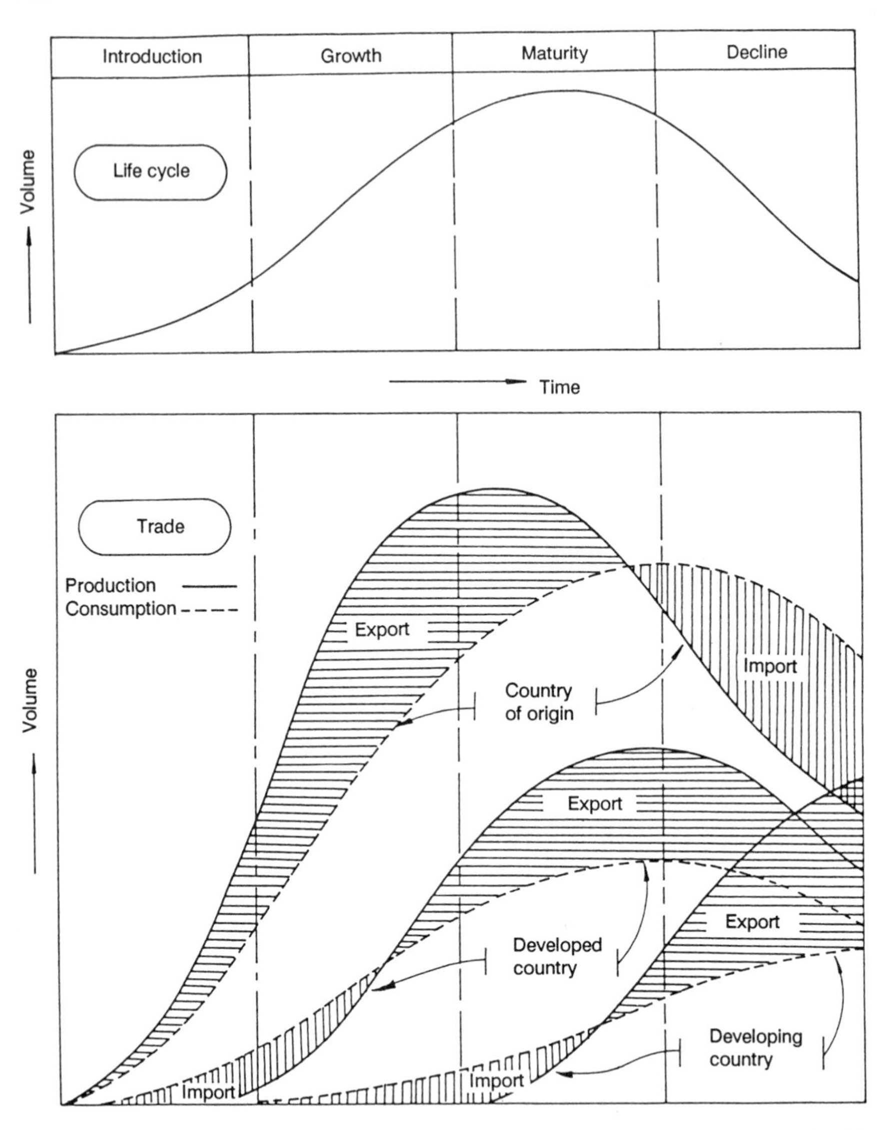

FIGURE 1 Technology life cycle and international trade. SOURCE: Economic and Social Commission for Asia and the Pacific (1984).

to decreased profit and less favorable terms of trade. New and old products also differ in the ratio of intellectual content to mass, with new products containing less mass and more skill. Even in material-based mass products such as cars and machines, this trend is evident in the microprocessors, chips, and plastics needed for the products to function. This trend has contributed to globalization by making it easier to break the production process into components and rendering materials and skill more portable.

The dynamics of technology flows, comparative advantage, and prosperity will then largely depend on the rate at which an undeveloped country can absorb new technologies. The role of protectionism as both stimulus and brake in this absorption process is critical. Methods used by almost every nation to protect infant industries include giving key industries preferred status and R&D support. Yet, if protection is sustained, it distorts the economy and may be the most serious impediment to growth through globalization of technologies.

Against these theoretical concepts, development in the Pacific Rim region illustrates a wide variety of starting points and outcomes. The region demonstrates shifts in comparative advantages; structural changes in the economies, particularly in manufacturing; problems of scale and fragmentation, cultural adaptability, and resistance to change; economic policies from free trade to central planning; and attitudes on the role of the transnational company in globalization ranging from "open arms" policies to controlled cooperation and even concealed or overt distrust.

The Mode of Globalization of Technology

The means of creating and transferring technologies deeply influence the impacts of those technologies. Identical methods produce vastly different effects in different environments. The classic mode—scientific publication—while still crucial on a worldwide basis, has receded in importance compared with rapid communication between peer groups in academe and transnational corporations.

The linear model—the direct path from science to economic development—no longer holds fully for the world. Science now owes at least as much to technology as technology owes to science. In developing countries the model is misleading, since so much of the locally developed science is doomed to lie idle because of the lack of development capacity. The result has been a vastly increased importance of the transnational company as the most effective means of generation and transfer of technology. Accelerating globalization by the breakup of the production process into elements and the unprecedented speed of transfer of complex data have enabled the transnational companies to assist developing nations in evolutionary jumps, bypassing whole stages. In the extreme case, this may produce technology without comprehension and economic advance with a high level of dependence. The opposite extreme is excessive reliance on local science as the principal source of local technology, which has produced imbalances between expenses on public sector science and inadequate benefits to private sector technology. Where the local science-push model failed, the result has been

TABLE 1 Chronology of Turning Points

Country or Region	Date
Europe, North America	1850
Malaysia	1850
Thailand	1850
Mexico	1876
Japan	1880
Taiwan	1895
Philippines	1900
Korea	1910
People's Republic of China	1949
Indonesia	1965

SOURCE: Reynolds (1985).

disenchantment with, and resentment of, dependence on international technology. This situation demonstrates that "inappropriate" and "appropriate" technologies are short-lived in relation to the cultural impact of "inappropriate" education and an "inappropriate" national ethos in science and technology.

Gross National Product and Growth Rates

Globalization of industries has caused rapid growth rates of the gross national product (GNP) in the Pacific region. This is largely because of a belated and accelerated entry into the industrial era, similar to but faster than earlier growth phases in Europe and North America. Reynolds (1985) has identified turning points in different nations and regions, when growth of per capita income began to exceed population growth and began real growth (see Table 1). His chronology of these points starts with a first boom from 1850 to 1914 and culminates in the second boom, the golden age of growth, from 1945 to 1973. Some remarkably early turning points in the Pacific region were largely due to primary product exports—sugar, rubber, and minerals—stimulated by industrialization in the West. Growth rate statistics for recent years reveal how globalization affected industries in the Pacific region (see Table 2 and Figures 2 and 3).

Japan, Oceania, the newly industrializing Asian nations (Hong Kong, Singapore, South Korea, and Taiwan), and the Association of Southeast Asian Nations (ASEAN) collectively grew by 6 percent, and their manufacturing industries grew by 7.2 percent per annum. In relative terms, the ASEAN nations, the Asian NICs, and Mexico have growth rates that exceed those of the advanced countries, although the base from which they began was low. By 1986 Japan had overtaken all advanced nations in rate of growth and all but the United States in per capita gross domestic product. Australia

TABLE 2 Gross National Product of Selected Countries, 1983

Country or Region		Value in 1980		Average Annual Growth (%)		Real GNP (billion U.S. dollars) in 1983	Per Capita GNP (U.S. dollars) in 1983
		1971	1983	1975/1971	1983/1975		
Advanced Countries		—	—	—	—	4,267.4	—
Australia	Billion Australia Dollars	95.4	131.5	3.3	2.4	145.7	9,474
Canada	Billion Canada Dollars	213.3	303.4	4.6	2.2	246.2	9,891
Japan	Billion Yen	152,972.0	262,073.0	4.7	4.6	1,103.4	9,252
New Zealand	Million New Zealand Dollars	20,382.0	25,484.0[a]	4.4	0.7[b]	33.9	10,725
United States	Billion U.S. Dollars	2,002.6	2,738.2	2.3	2.8	2,738.2	11,677
Asian NICs		—	—	—	—	130.9	—
Hong Kong	Million Hong Kong Dollars	59,921.0	178,071.0	6.4	11.1	24.5	4,623
South Korea	Billion Won	18,770.0	46,734.0	8.9	7.4	58.0	1,450
Taiwan[c]	Million Taiwan Dollars	738,712.0	938,877.0	7.7	8.7	48.4	2,602
ASEAN Countries		—	—	—	—	—	—
Indonesia	Billion Rupiahs	22,561.0	52,253.0	8.3	6.7	57.5	367
Malaysia	Million Ringgit	25,725.0	62,143.0	7.5	7.7	26.8	1,799
Philippines	Billion Pesos	151.4	282.6	6.7	4.7	25.4	488
Singapore	Million Singapore Dollars	11,404.0	28,393.0	8.8	7.5	13.4	5,360
Thailand	Billion Baht	367.4	802.1	6.7	6.7	34.9	705
Island Countries		—	—	—	—	—	—
Fiji	Million Fiji Dollars	726.2	990.6	5.3	1.3	—	—
Papua New Guinea	Million Kina	1,309.0	1,735.0	6.0	0.6	1.4	417
Latin America		—	—	—	—	99.6	—
People's Republic of China	Billion Yuans	91.9	103.0[a]	Δ0.1	1.7[b]	54.4	214

[a]Values in 1982.
[b]Values in 1982/1975.
[c]Value in 1981.
SOURCE: PBEC Japan Member Committee of the Tokyo Chamber of Commerce and Industry, *Pacific Economic Community Statistics, 1986*. (Tokyo: Tokyo Chamber of Commerce and Industry, International Division, p. 27, 1986.)

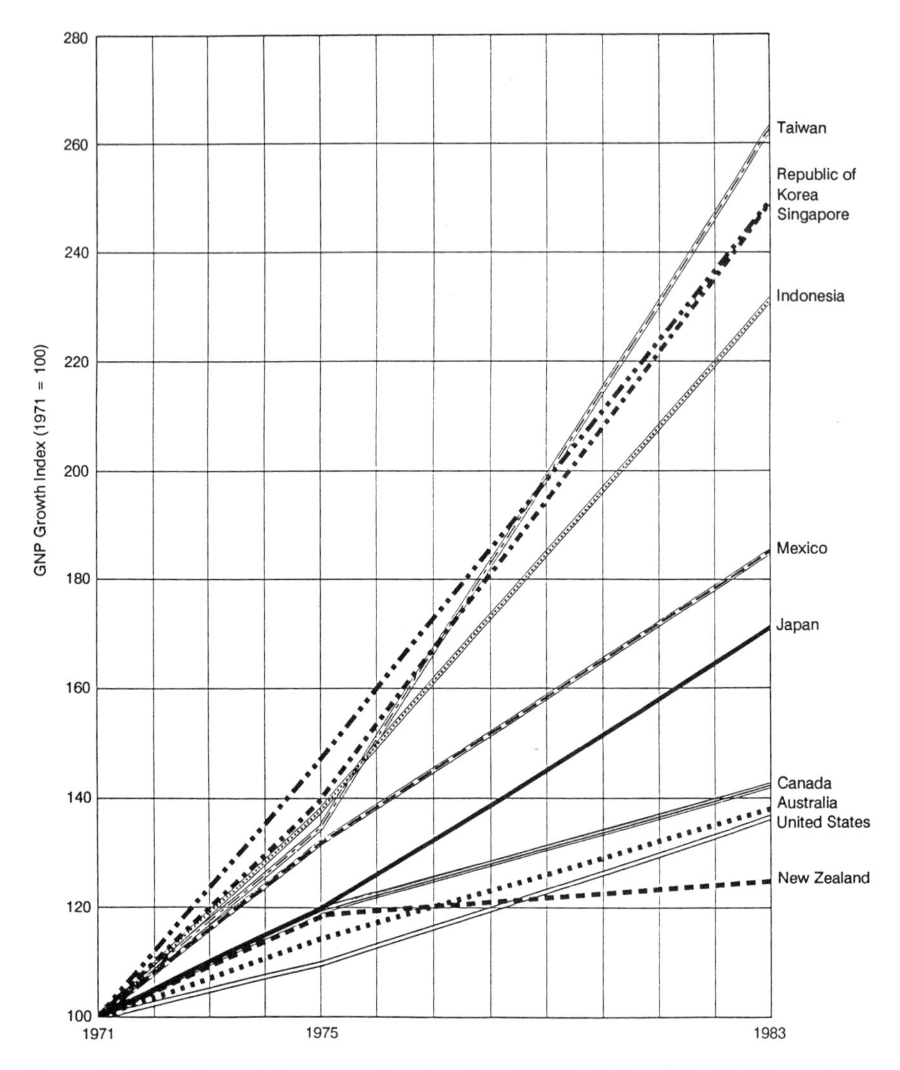

FIGURE 2 Rate of growth in gross national product (GNP) of selected Pacific Rim nations, 1971–1983. SOURCE: World Bank (1985).

and New Zealand were intermediate but have dropped back in recent years. The Asian NICs and the ASEAN nations, with the exception of the Philippines, displayed striking resilience during the two postwar oil shocks.

GNP growth forecasts, usually based on sophisticated computer models, suggest attenuated continuation of these trends (Findlay et al., 1986; Onishi and Nakamura, 1986). Hence, the mean growth rates projected by these

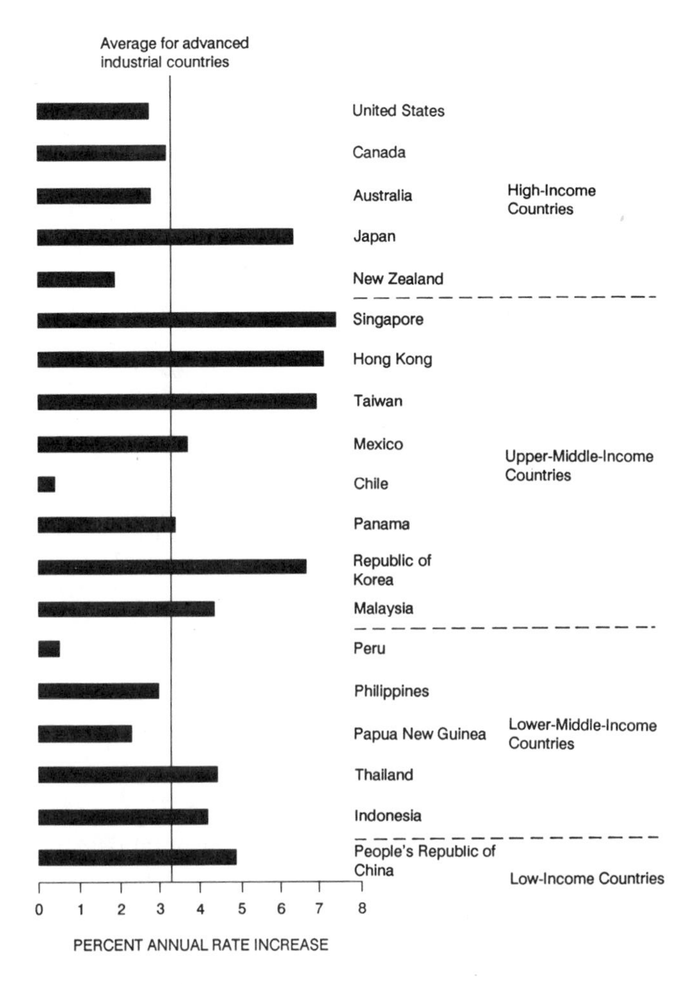

FIGURE 3 Mean rate of increase in per capita gross domestic product of selected Pacific Rim nations, 1960–1982. SOURCE: World Bank (1984).

authors for the period 1980 to 1995 are about 5.6 percent for the Asian NICs, 4.7 percent for the ASEAN nations, 4.1 percent for Japan, and 3.4 percent for Australia. For the fully industrialized countries of North America and Europe, growth rates are projected at 2.3 percent. Low rates are projected for the Philippines, and much faster growth from a low base is projected for the People's Republic of China (see Table 3).

Globalization and Structural Change

Some fruits of the Industrial Revolution were introduced into the Pacific region in the second half of the nineteenth century by the colonial powers,

TABLE 3 GNP Growth Rates, Actual and Projected, 1962–1981, 1990, 1980–1995 (percent)

	Growth Rate			
Region or Country	Actual[a] (1962–1970)	Actual[a] (1970–1981)	Projected[a] (1981–1990)	Projected[b] (1980–1995)
World	5.7	3.6	3.4	3–3.3
Pacific Basin	—	—	—	3.8–3.6
People's Republic of China	5.7	2.8	9.0	8.4–5.9
Japan	14.9	5.3	3.8	4–4.2
Other Northeast Asia	3.2	8.4	8.5	—
ASEAN	12.6	7.4	6.0	3.5–4.5
North America	4.1	3.1	3.5	2.7
Australia	5.8	3.5	3.8	3–2.5
New Zealand and other Pacific	3.7	3.6	3.0	1.9–1.2
Europe	7.4	2.1	2.1	2.6–2.8
Middle East	8.9	8.5	4.1	1.3–3.5
Hong Kong	—	—	—	5.5
South Korea	—	—	—	7–5.6
Singapore	—	—	—	4.2–5.8
Indonesia	—	—	—	3.7–4.1
Malaysia	—	—	—	4.6–4.2
Philippines	—	—	—	0–2.9
Thailand	—	—	—	5–5.1

[a]SOURCE: Adapted from Findlay et al. (1986).
[b]SOURCE: Adapted from Onishi and Nakamura (1986).

merchants, and migrants or, in Japan, by government initiative. These were mainly agricultural, extractive, and transport technologies. Exports of primary products led early growth. Those activities produced trading surpluses to support the gradual buildup of protected import replacement industries.

After World War II, the region was influenced by the United States, which had reached a peak of economic expansion, and Japan, which was undergoing the transformation into the second-largest economic force of the industrialized economies. Factors contributing to the expansion of Japan and the United States and to their subsequent influence in the region were local raw materials from Australia and Indonesia for Japan, technology transfer, capital, and cheap labor, which encouraged local investment by both nations.

As people, products, knowledge, and capital became more mobile, the effects of the product cycle accelerated. Production costs became more sensitive to the shifting comparative advantage, and the progressive lowering of trade barriers facilitated transfers. This trend was particularly pronounced in the Pacific because of large differences in labor and raw materials. Industrialization began with simple manufacturing of consumer goods and processing of local raw materials. In recent years, the breakup of the production process into standardized segments promoted relocation of labor-intensive steps into the countries with the most favorable production environment. Whereas classic trade in finished goods centered on the developed nations, this new trade in components linked to internationally integrated production lines favored and stimulated the developing countries.

Within these broad generalizations about development and trade, several regional patterns have emerged. The Asian NICs initially concentrated on labor-intensive light industrial goods but moved to more advanced technologies stimulated, for example, by Singapore's minimum wage regime. These goods were increasingly skill-based, but in South Korea and Taiwan this included capital-intensive processes used to produce petrochemicals and steel and capital goods used in heavy engineering. The Philippines, Malaysia, and Indonesia, while continuing to rely heavily on agriculture, followed the NICs with light industries as well as some capital-intensive investments. Australia, with rich resources and high labor costs, coped with declining terms of trade for resource-based products by expanding production and improving productivity in agricultural produce and coal. However, Australia's import replacement industries declined, and so far the nation has not benefited greatly from endeavors to participate in integrated international productions, e.g., in the automobile industry. The classic example of Japan, although well-known, bears restating here because it has served as a model to many developing nations in the region (Sekiguchi and Horiuchi, 1984). Since the Meiji Restoration, which began in 1868, Japan has been transformed from an economy based on agriculture and textiles to an industrial power of the first order.

The stages of change can be summarized briefly. First came a long period of protection and promotion of almost every light industry aimed at absorbing the labor force. Essentially, only light industries were considered appropriate at the time, in view of the abundance of cheap labor; they provided the exports to finance raw materials and food. After World War II, Japan adopted the policy of priority production and allocated resources preferentially to basic sectors—coal, steel, and electricity. In the 1960s, Japan shifted to heavy manufacturing industries and agricultural diversification, away from rice as the sole product. Local infant industries—electrical machinery, motorcycles, cars, planes, petrochemicals, and electronics—were subsidized, protected against imports, and encouraged to export.

In 1960, Japan took the first step of liberalization by changing from quantitative restrictions to tariffs in its Program of Liberalization of Trade and Foreign Exchange. The program was followed in 1963 with the acceptance of Article 11 status under the General Agreement on Tariffs and Trade, by substantial liberalization of imports and by partial relaxation of control of foreign investments. By the time Japan liberalized foreign investment completely, the basic industries were fully established, and the country had a vast foreign exchange surplus.

Japan continued to lead in government-inspired structural change. By the

TABLE 4 Added Value in Agriculture and Manufacturing (constant prices, local currency converted to U.S. dollars)

Country	Period	Compound Growth Rate Per Annum		Employment in Manufacturing (1970–1980 compound growth per annum)
		Agriculture	Manufacturing	
Australia	1970–1982	+0.51	+0.47	−1.1
Canada	1970–1983	+2.6	+0.94	+1.24
Hong Kong	1970–1982	−3.6	+4.8	—
Indonesia	1970–1984	−3.4	+8.4	+1.03
Japan	1970–1982	+3.4	+10.3	−0.6
South Korea	1970–1984	−3.8	+6.4	+8.9
Malaysia	1970–1984	+6.6	+16.5	—
Mexico	1970–1984	−14.3	−10.4[a]	−7.8[a]
Philippines	1970–1984	−3.6	−0.3	—
Singapore	1970–1984	+4.6	−11.6	+8.77
Thailand	1970–1984	+3.1	+7.9	—
United States	1970–1983	+0.34	+3.3	+0.53

[a]Currency effects.

SOURCE: World Bank world tables, provided and compiled by Corinne Boyles, Australia-Japan Research Centre, Australian National University, Canberra, Australia.

1970s rising labor costs and environmental constraints in an overpopulated country made further emphasis on heavy industries less desirable. The Ministry of International Trade and Industry (MITI) introduced its "knowledge-intensive industrialization" strategy, which put a government policy stamp on the trend initiated by the private sector in the United States and later followed by all and sundry, with the catch-phrase "sunrise industry." Yet large residues of protectionism for farmers—rice and grain—and a variety of nontariff devices remained and are being reversed only slowly in the 1980s. With economic success on almost all fronts in this decade, the liberalization that prosperity can afford is occurring. Government administrative guidance is being questioned and, following the U.S. example, original research and innovation are seen as the final driving force of industrialization.

Structural change in the other Pacific countries has been less dramatic than that in Japan and is at different stages, but it is nevertheless somewhat similar. The ways in which globalizing industries such as fertilizers, agricultural machinery, and pesticides affect agriculture have been similar in most countries; intensification and increased output and productivity have resulted in declines in employment. However, terms of trade in agricultural produce have deteriorated worldwide, and hence, growth in added value was smaller than that in manufacturing or was even negative (see Table 4).

Throughout the Pacific region, exports have been particularly important and closely linked to growth rates. Exports provide links to the developed nations and therefore accelerate learning curves and induce competitiveness while overcoming two of the major impediments to growth in small nations—the disadvantage of scale and the shortage of foreign exchange. Exports also demonstrate the areas of local economic strength and are therefore a fairly sensitive indicator of comparative advantage and how it shifts with time. Kuznets (1984) has used the revealed comparative advantage (the ratio of a country's share in world exports of a particular product to its share in world exports of all manufactures) as an indicator of shifts in comparative advantage and, hence, structure (see Table 5). Thus, in the United States where pharmaceuticals are a research-intensive high-technology industry, the industry has sustained and improved its comparative advantage; by contrast Hong Kong's trade in generic drugs—the formulation of drugs no longer protected by patent—is essentially low technology and has declined. In wood products, countries with the advantage of natural raw materials have performed well, particularly if they also have low labor costs. Taiwan, South Korea, and Malaysia have done well, but with rising labor costs, their advantage declined. Textiles and toys demonstrate Japan's progressive disengagement in favor of the NICs. In office machines, the technical superiority of the United States has been maintained in absolute terms against Hong Kong's components and Japan's faster rise, particularly since 1978. The emergence of the NICs is notable in the manufactures that are technology- and labor-intensive, such as telecommunications equipment, toys, and sporting goods. The lighter labor-intensive industries—apparel, toys, sporting goods—have remained a domain of Hong Kong. In technology- and capital-intensive mass products (motor vehicles) and scientific instruments, the United States and Japan continue to dominate. In motor vehicle production, there has been a rapid relative strengthening of Japan and a sharp decline of Australia.

Another way of tracing structural change through trade patterns is to examine the ratio of imports to exports (see Table 6), that is, the dependence on imports, the reverse of comparative advantage. A comparison of South Korea and a prosperous developing country—Malaysia—with the United States and Japan shows the trends: The United States and Japan show growing dependence on raw and refined materials with increasing industrial production, whereas South Korea and Malaysia show declining dependence on refined materials and imported capital goods such as machinery. On the whole, the picture is similar to the related indicator of revealed comparative advantage.

In the light of the shift of comparative advantage, the temptation is great to see real progress in this pattern—development from the lower to the next stage of industrialization—at least for some time to come. Although such an optimistic picture is the opposite of the ever-widening gap between North and South predicted by some economists, the Pacific region offers some basis for optimism. It has a suitable gradation of stages of development, ample

TABLE 5 Comparative Advantage by Industry Group, 1970s

Industry and Country	1970	1978	Industry and Country	1970	1978
Drugs (SIC 541)			Motor Vehicles (SIC 732)	1.26	0.40
Hong Kong	1.24	0.63	Australia	3.83	3.87
Japan	0.24	0.17	Canada	0.88	1.62
Singapore	1.11	1.65	Japan	0.79	0.26
United States	0.99	1.23	Singapore	1.08	1.16
Wood Products (SIC 631, 2)			United States		
Canada	2.24	3.43	Apparel (SIC 84)	9.49	10.13
Japan	0.90	0.19	Hong Kong	0.86	0.17
South Korea	17.65	5.80	Japan	10.67	7.50
Malaysia	26.14	8.23	South Korea	4.12	4.00
Taiwan	8.46	4.34	Taiwan		
United States	0.60	0.75	Scientific, Other Instruments (SIC 681)	0.14	0.36
Textiles (SIC 65)			Canada	0.34	0.85
Hong Kong	1.71	1.92	Hong Kong	1.45	1.91
Japan	2.46	0.87	Japan	1.64	1.21
South Korea	1.95	2.91	United States		
Taiwan	1.67	1.55	Toys, Sporting Goods (SIC 894)	9.90	10.26
Office Machines (SIC 714)			Hong Kong	2.16	0.89
Canada	0.75	1.23	Japan	0.48	3.62
Hong Kong	0.11	1.26	South Korea	2.57	2.73
Japan	0.76	0.88	Taiwan	0.70	1.13
Taiwan	0.73	0.51	United States		
United States	2.32	2.63			
Telecommunications Equipment (SIC 724)					
Hong Kong	1.60	1.92			
Japan	2.89	2.48			
South Korea	0.32	1.93			
Taiwan	3.69	2.37			
United States	0.90	0.98			

NOTE: SIC = Standard Industrial Classification Code.
SOURCE OF DATA: United Nations, *Yearbook of International Trade Statistics* (New York: United Nations, 1971, 1972, 1978, 1979).
SOURCE: Adapted from Kuznets (1984, pp. 50–51).

TABLE 6 Historical Trade Patterns of Selected Countries, Import/Export Ratio

	Type of Production					
Country	Raw Materials	Refined Materials	Processed Materials	Manufactured Goods	Machinery	Total
United States						
1972	1.28	1.05	1.46	2.08	0.82	1.16
1975	2.22	0.58	0.93	1.54	0.54	0.90
1979	2.85	0.74	1.13	2.21	0.77	1.23
Japan						
1972	23.29	3.70	0.34	0.33	0.18	0.82
1975	22.57	12.62	0.27	0.46	0.14	1.03
1979	42.79	13.81	0.43	0.66	0.15	1.07
South Korea						
1972	4.90	4.70	1.13	0.13	4.42	1.55
1975	9.82	1.51	1.06	0.10	2.68	1.43
1979	18.39	1.38	0.89	0.14	1.90	1.34
Malaysia						
1971	0.29	1.14	0.98	4.03	9.57	0.80
1974	0.23	0.92	1.34	1.42	7.11	0.97
1979	0.22	0.55	1.37	1.32	2.39	0.76

SOURCE: Economic and Social Commission for Asia and the Pacific (1984, p. 105).

raw material resources—with the possible exception of liquid fuels—and the necessary human resources. All the nations have demonstrated that they can cope with technological change—something that cannot be asserted with equal confidence in other parts of the world. Indeed, despite earlier revolutionary rumblings in Malaysia and Indonesia, rising prosperity has given the area remarkable political stability compared with that in parts of South America and Africa. The trend toward prosperity is continuing.

Even with the rosy spectacles of this generalization, the question arises of just how far the complementarity of the different countries goes and what happens when all are highly industrialized and skill based. Obviously, there will be no definitive answer to this question for years. But some of the likely problems can be illuminated by a look at countries that are representative of both a particular problem and a particular group of countries in the region.

ASEAN AND COMPLEMENTARITY: GROWTH IN AGRICULTURE AND MANUFACTURING

Strengths and Characteristics

The strengths of the ASEAN economies lie in their raw materials, cheap labor, and the export-oriented strategy of all five countries,[3] particularly that of Singapore. These strengths have helped them to overcome the disadvantages of scale. Except for Singapore, their economies and employment are still largely based on agriculture, with their main products being rice, maize, rubber, timber, sugar, and crude oil. However, industrialization is proceeding quickly, and manufactured goods have reached 20 percent of the gross domestic product (GDP). This figure is the crossover point, at which some nations begin a decline in manufacturing and a transition to service- and knowledge-based postindustrial structures.

Common characteristics of the ASEAN economies are as follows:

- The ASEAN countries have adopted—to varying degrees—open economies that have led to increasing external trade, foreign investment, and technology transfer.
- As a result of natural endowments and political stability, the area has grown faster than most developing countries.
- ASEAN external exports have grown 43 percent faster than world exports and doubled as a percentage of world exports, from 1.76 to 3.6 percent.
- The engine behind ASEAN's improving external trade has been the United States; U.S. imports from ASEAN have grown 16 percent faster and exports to ASEAN have grown 12 percent slower than U.S. trade with the rest of the world.
- Despite growth in manufacturing, Japanese and U.S. investment in ASEAN has been strongest in agriculture and extractive industries (Buchanan, 1986).

Problems and Challenges

The structure of the five ASEAN economies is such that they are inherently more competitive than complementary. Hence, prosperity depends largely on the external engines of growth, the United States and Japan. The policy of import substitution is likely to aggravate this situation. However, ASEAN trade with PEC countries other than the United States and Japan has grown faster than has trade with the latter. Despite inter-ASEAN Preferential Trading Arrangements, substantial tariff barriers of up to 50 percent persist. All ASEAN governments have ambitious industrialization plans that will be difficult to harmonize and that will, in turn, eventually compete with those of South Korea and Taiwan. Unless demand rises with increased production, industrialization in the region will be like "musical chairs," as each country expects the countries on the next rung up on the industrialization ladder to vacate some of their seats.

ASEAN is also facing problems in the agricultural sector. Prices of primary commodities have declined by 1.25 percent per annum since 1900 (Drucker, 1986; Inoguchi, 1986; World Bank, 1986). In the Philippines in particular, where about half of the exports are in agriculture, low commodity prices and high interest rates have affected their industrialization program. This, in turn, may well affect political stability—perhaps the weak spot in the region. The problem is aggravated by South Korea's and Japan's tight price protection of agricultural products and bilateral deals, as well as price wars between the EEC and the United States. The result of these difficulties is a serious trade imbalance in the Philippines.

In the sociopolitical sphere—despite steady improvements—uncertainties remain in some ASEAN nations. In the Philippines, large land areas and cultural privilege are still the preserve of a small proportion of the population. Where economic, educational, and cultural privilege is held by a minority, economic flexibility and mobility are impeded, and the emerging intellectual proletariat tends to be radicalized. In Indonesia, geographic fragmentation has created problems in organization and education. These problems, combined with the desire for rapid national development, tend to give rise to "strong-hand" military-type governments. To what extent these political characteristics will affect stability and economic success remains unknown.

Finally, ASEAN nations—perhaps with the exception of Singapore—are not yet at the stage where their science can make significant original contributions to indigenous technology. Attempts to develop "appropriate" technologies by local scientists have been problematic. At best, such developments are reverse engineering of earlier technologies; often they are difficult to time in view of rapid alternative developments and tend to prolong uncompetitiveness. The problem of indigenous science and its conversion into technology in small economies is dealt with later in this paper in regard to Australia, where the problem has become more pronounced.

SOUTH KOREA: TECHNOLOGY FIRST, EDUCATION SECOND, SCIENCE THIRD

If a rapid rise in living standards from a low level is the prime objective of globalization of industries, the Asian NICs demonstrate the effectiveness of globalization. These four countries have led in international GDP growth rates and have contributed about half of all manufacturing exports from the Third World (Reynolds, 1985). Although politicians and economists blamed many of the shortcomings of Third World industrialization on the advanced nations, particularly their multinational corporations (MNCs), it was by interaction with these corporations through direct foreign investment, technology transfer, importation of capital goods, joint ventures, and competition in the open market that the four Asian NICs have advanced rapidly. Some of the factors contributing to their rise will be discussed here, referring to South Korea only for simplicity's sake, since it is typical, although not wholly representative, of the group.

Strategies for Growth

Compared with the relative advantages of the established international ports of Hong Kong and Singapore, South Korea rose from a particularly low base. It had experienced the partition of its economy into two unequal halves and the national disaster of a fratricidal war in 1950–1951. But South Korea catapulted to the status of an industrializing and highly competitive nation in one generation, following the path of Japan more quickly than any other country. The economy was essentially a form of planned, nationalistic capitalism which may well have been a reaction to tight domination by Korea's earlier colonial masters, the Japanese. In the late 1930s the Japanese owned 62 percent of all landholdings of over 250 acres and some 90 percent of industrial capital (Reynolds, 1985). From the outset in the 1950s, technology transfer aimed at high levels of South Korean ownership. The ability to borrow was helped greatly by injections of American capital, which were made largely but not solely for strategic reasons. During the early years of strict government supervision, policies on licensing and investment were deliberately kept separate and distinct (Kim, 1984). Licenses were carefully supervised and were tied to export promotion, capital goods intermediates, and demonstrable diffusion effects to other sectors. Royalties were capped by a 3 percent ceiling, and agreement contracts were confined to 3 years. Naturally, only mature and more labor-intensive technologies could be licensed under these terms.

The first liberalization of royalty ceilings and contract periods and a reduction in red tape occurred in 1970. These changes broadened the scope for licenses in keeping with South Korea's rapid advance in technological sophistication. In 1984, when local industry was viewed as sufficiently established but not sufficiently competitive, licensing policy was opened to all forms of industry and agreements. Direct foreign investment policy, by contrast, was liberal from the outset in 1960 and allowed all forms of investment, including wholly owned subsidiaries and access to extensive incentives. A reversal of this policy occurred in 1973, to prevent adverse effects on local firms. Joint ventures with local companies were given preference; export requirements were imposed, but exceptions were made for high technology and 100 percent export enterprises. Finally, in 1984, direct foreign investment policy was again made more liberal, with some two-thirds of all industries being approved automatically, particularly the high-technology industries. Full liberalization to the level of the advanced countries is planned for the current Fifth Five-Year Plan (1984–1989). Thus, for a long time South Korean policy was far from an open economy or an MNC ideal. Yet South Korea is now close to an open, outward-oriented economy, has a high level of technological self-confidence, and in many ways has advanced beyond the status of an industrializing nation. It is a major exporter of manufactures,

TABLE 7 Value of Foreign Investment by Ownership Pattern in South Korea, 1962–1980 (in $U.S. million)

Industry	0–49% Value	50% Value	51–99% Value	100% Value	Total Value
Foods	6.2	9.5	10.0	1.9	27.6
Textiles and apparel	5.9	29.3	36.0	2.1	73.4
Wood and paper products	0.2	0.0	—	0.9	1.1
Chemicals and petroleum	52.9	199.5	3.1	102.4	358.0
Nonmetallic mineral	6.0	—	3.9	3.5	13.5
Metals	16.7	11.2	20.3	11.3	59.5
Machinery	38.0	26.0	5.4	27.5	96.9
Electrical machinery	45.6	32.8	16.0	77.1	171.5
Transportation equipment	1.2	29.9	5.8	0.8	47.8
Others	21.0	3.7	3.2	5.8	33.7
TOTAL	193.7	351.9	103.7	233.3	883.0
By country[a]					
United States	82.6	117.2	8.5	72.2	280.5
Japan	109.8	192.2	121.1	217.8	641.0
Others	91.8	120.0	9.1	79.5	300.3
TOTAL	284.2	429.4	138.7	369.5	1,221.8

[a]Country data include nonmanufacturing investments.
SOURCE: Economic Planning Board. 1981. Foreign Direct Investment Special Survey. Seoul, South Korea.

and even of technology (mainly to less developed countries), and an effective partner of MNCs in international integration of sophisticated product lines.

All this was achieved with a remarkably low proportion of foreign ownership. By 1980 capital imports had reached $19.2 billion, of which less than 6 percent, $1.1 billion, was foreign direct investment, mainly in manufacturing (Park, 1984) (see Tables 7, 8, and 9). However, this is not the whole story. Capital goods imports from the United States and Japan were 14 and 20 times, respectively, as much as all other forms of technology transfer combined. In contrast, 64 percent of South Korea's general machinery and 67 percent of its electrical machinery exports to the United States came from American MNCs in South Korea (Helleiner, 1979). The pattern of foreign investment reflects the changes in government policy from an emphasis on labor-intensive to labor/technology-intensive and, eventually, technology-intensive industries. Thus, for example, investment in textiles virtually stopped and that in electrical machinery and electronics accelerated in later years, while that in chemicals and transport peaked in the mid-1970s.[4]

If acquisition of technology was South Korea's first objective, education was its second; it was closely timed and balanced with the country's technology policy. Expenditures on education rose from 2.5 percent in 1952 to 21.9 percent in 1984, but represented only 30 percent of the total cost, with the balance being provided by the private sector. By 1980, enrollment rates of those in the appropriate age groups were more than 100 percent (including older students) in elementary schools, 70 percent in high schools, and almost 25 percent in universities. Measured on a GDP per capita/education correlation graph (Harbison and Myers, 1964), South Korea was two to four times higher up the scale than predicted, ahead of Mexico, Argentina, and Brazil.

Science R&D was very much a third priority, well behind engineering; between 1972 and 1980, the science budget increased from only 0.31 percent to 0.95 percent of GDP. However, as in Australia, endeavors to strengthen the science supply were ineffective (Kim, 1984), as long as there was no real competitive demand for indigenous innovation in the economy. The link remained weak until the late 1970s, when scientists were involved in the evaluation and transfer of technology, and when measures were taken to create a more competitive market environment. These measures include further liberalization of trade through removal of all entry barriers, tariff reductions, and reduced subsidies for targeted strategic industries and were timed with a 20 percent annual increase in R&D to reach 2 percent of the GDP by 1986. In the future, special emphasis will be placed on higher

TABLE 8 Value of Foreign Investment by Year and by Industry in South Korea, 1962–1980 (in $U.S. million)

Industry	1962–1971 Value	1972–1976 Value	1977–1980 Value	Total Value
Foods	1.2	1.7	24.7	27.6
Textiles and apparel	13.4	57.8	2.2	73.4
Wood and paper products	—	0.3	0.8	1.1
Chemicals and petroleum	42.9	185.0	130.1	358.0
Nonmetallic mineral	5.6	4.4	3.5	13.5
Metals	9.0	30.4	20.1	59.5
Machinery	5.3	31.7	59.9	96.9
Electrical machinery and electronics	12.0	84.5	75.0	171.5
Transportation equipment	0.0	27.3	20.5	47.8
Others	2.4	8.4	22.9	33.7
TOTAL	91.8	431.4	359.7	883.0

SOURCE: Economic Planning Board. 1981. Foreign Direct Investment Special Survey. Seoul, South Korea.

TABLE 9 Technology Transfer to South Korea from the United States and Japan, 1962–1984

Transfer	Value (in thousands of dollars)					
	1962–1966	1967–1971	1972–1976	1977–1981	1982–1984	Total
Foreign licensings						
U.S.	533.1	7,816.2	21,265.5	159,156.1	185,994.0	374,784.9
	(71.2)	(48.1)	(22.0)	(35.3)	(52.5)	(40.8)
Japan	—	5,041.6	58,653.5	139,808.1	92,874.3	296,377.5
		(31.0)	(60.8)	(31.0)	(26.2)	(32.3)
Direct foreign investments						
U.S.	21,871	12,443	67,924	208,097	298,103	608,438
	(95.1)	(17.1)	(12.0)	(35.4)	(38.7)	(30.1)
Japan	693	40,759	376,940	214,906	341,085	974,383
	(3.0)	(56.1)	(66.7)	(36.6)	(44.3)	(48.3)
Capital goods imports						
U.S.	75,249	472,438	1,972,631	6,218,961	5,627,422	14,366,701
	(23.8)	(18.6)	(22.3)	(22.2)	(28.6)	(24.2)
Japan	147,987	1,291,995	4,423,596	14,268,593	7,576,106	27,708,277
	(46.8)	(50.9)	(50.0)	(51.0)	(38.5)	(46.7)
Technical consultancy						
U.S.	—	12,137[a]	7,674	20,780	8,747[b]	49,338
		(72.3)	(41.5)	(38.0)	(44.6)	(45.0)
Japan	—	3,133[a]	5,984	16,704	6,214[b]	32,035
		(18.7)	(32.4)	(30.5)	(31.7)	(29.2)

NOTE: Figures in parentheses indicate the percentage of total transfer.

[a]Only 1968–1976 data are available.

[b]Only 1982 data are available.

SOURCE: L. Kim. 1984. Technology Transfer and R&D in Korea: National Policies and the U.S.–Korea Link. Paper presented at the Conference for National Policies for Technology Transfer, Hawaii, October 8.

education, primarily in engineering and science, where target figures are 6,000 and 1,500 Ph.D.'s, respectively. Several thousand South Koreans will continue to be trained overseas, particularly in the United States, as has been the practice for years.

Problems and Challenges

Latecomers to the manufacturing sector, particularly to the capital- and technology-intensive sectors, lack comparative advantage in technology. They tend to retain their protection mechanisms, while also requesting access to the major free markets. It is not entirely surprising that in response the United States has intermittently resorted to emergency relief protection. Hufbauer (1970) claims that imports covered by special protection in the United States (at that time) still equaled $68 billion, or 21 percent of total U.S. imports. Although South Korea has been treated generously by the United States, South Koreans believe that this is due compensation for their frontline defense of the West. Protectionist measures by the United States have been a sensitive issue.

As the closest follower of Japan, South Korea is the first Asian NIC to compete with Japan in the production of technology-intensive capital goods. With rising living standards, the comparative labor advantage will decline, although it will remain substantial for some time. Singapore has already experienced the phenomenon of "lowered motivation," that is, the younger generation's unwillingness to work the long hours at low wages, which their parents accepted in order to achieve prosperity and security.

The transfer of technology to South Korea may face obstacles in the future. South Korea does not have a particularly good reputation with regard to its respect for the copyright of software. Japan is already reluctant to transfer high technology, and the United States will increasingly become so as South Korean competition approaches U.S. capacity.

South Korea's step onto the highest rung of technology will depend on the effectiveness of its current preparations for more indigenous South Korean R&D. In relying on their own innovations, the South Koreans and other Asian NICs will increasingly meet R&D problems that Australia is already facing.

AUSTRALIA: NATURAL RESOURCES FIRST, SCIENCE SECOND, TECHNOLOGY THIRD

Australia has a high standard of living, education, and science. Although its resource industries are highly developed, their terms of trade have deteriorated, and technology- and skill-intensive industries are intermediate between those of the NICs and the developed countries. Australia's ability to share in, and compete with, the globalization of these industries will depend largely on its own ability to convert science into technology and on its interaction with the world's multinationals. Other nations in the Pacific and elsewhere will face this issue as they lose their comparative advantage of low labor costs. The following section focuses on the cost-of-labor aspect of Australia's development.

Resource Development and Technology Transfer

Australia contributed significantly to the worldwide transfer of technology in the nineteenth century. Dynamite and gelignite for the gold mining industry were produced in Australia within months of the first manufacture by Nobel in the United Kingdom. Fertilizers were produced in the early 1870s. Australia had a multinational pharmaceutical company—Nicholas/ASPRO—in the 1920s, before many of today's pharmaceutical giants had become multinationals. The first two examples foreshadowed Australia's chief areas of development—mining and agriculture; pharmaceuticals did not succeed internationally. None of Australia's manufacturing companies attained significant multinational status.

International mining technology was transferred to Australia by early immigrants from the United Kingdom, the United States, and Europe and through Scottish and U.S. schools of mining, which have influenced their Australian counterparts since 1850. A series of mining booms during that period and new developments after World War II—nickel, iron ore, and gas in western Australia, gas and brown coal in Victoria, and coal in New South Wales and Queensland—attracted Australian entrepreneurs and many international companies, mainly from the United Kingdom and the United States and, more recently, Japan. As a result, a strong Australian mining industry developed, with companies with worldwide stature, such as Broken Hill Proprietary Co. Limited (BHP), Western Mining Corporation Limited, and Broken Hill North Proprietary Limited (B.H. North), and many joint ventures with international companies, including Mount Isa Mining Holdings Limited (MIM), CRA (formerly Conzinc Riotinto of Australia Limited), and Comalco Aluminum Limited (COMALCO). The large scope of many of the Australian deposits and the size of the companies involved ensured that Australia fully shared in and contributed to international mining. Two examples of Australian contributions illustrate this point—the development of flotation technology by Potter, Delprat, and de Bavay and Imperial Chemical Industries (ICI) Australia's contribution jointly with Cook-Farnham, Canadian Industries Limited (CIL) of Canada, and ICI U.K. to the large-scale development of ammonium nitrate slurries and water gels as safe explosives in underground and surface mining.

Most mining technology worldwide developed in steps that were stimulated by advances in engineering. Thus, technology was frequently spread through new mining equipment developed in the United States, the United Kingdom, and the Federal Republic of Germany. Using advanced technology developed by Exxon, Shell, and British Petroleum, Australian joint ventures with these and other multinationals drilled some of the world's deepest wells in the stormy seas off Australia's northwest coast and in the Bass Strait. Mineral discoveries have made Australia energy rich. Australia is virtually self-sufficient in oil (but not for long—only 50 percent by 1996 to 1997). Australia supplies Japan with vast and increasing quantities of liquefied natural gas, holds 19 percent of the world's uranium resources, and has dramatically increased its production of black coal to become the world's largest exporter and fifth-largest producer of coal (from 30 million tons in 1960 to 163 million tons in 1986). In preparation for a decline in oil resources, a joint Japanese-Australian venture in Victoria has built a $300-million brown coal liquefaction pilot plant. However, apart from supplies for domestic use, only a small proportion of Australia's ore is upgraded in Australia and traded with added value in refined form.

Australia is also richly endowed with agricultural resources, although in a form differing from that in most rural economies. Her wealth resides in her vast areas of land. Much of this land, however, is harsh and relatively infertile. These conditions greatly stimulated the application of local and international technology and science. Mechanization of agriculture and the use of fertilizer and pest control were absolutely vital for productivity. Australian farmers, supported by a large government technical service, became highly adept and flexible in their use of agricultural technology. International developments in machinery and chemicals were adapted rapidly. The Commonwealth Scientific and Industrial Research Organization (CSIRO) made many valuable contributions, such as the extermination of the rabbit and cactus plagues, the discovery of the importance of trace metals, and many incremental contributions, including optimal use of the biological control and growth agents that the multinational companies developed overseas and in Australia.

The dual wealth in mineral and agricultural resources made Australia an early island of Western prosperity in the Pacific. Prosperity and the scarcity of labor, particularly in periods of mining booms, at times catapulted Australian wages to levels well above those in Europe and in the process deprived Australia of the comparative advantage of competitive labor. This, combined with the disadvantages of a small market and long transport lines, operated against the growth of export-oriented manufacturing industries. Australia thus stood in contrast with other areas of the Pacific, where cheap labor was the basis of export and, hence, scale benefits. After World War II, Australia's policies were dominated by the need to populate her vast territory. Although agriculture was prosperous, it could not employ vast numbers of workers.

Industrialization in the cities was the only alternative: Australia changed from the "emptiest" continent to one with highly urbanized coastal areas.

Industrialization Policies

The Australian government pursued a two-pronged policy that began in the 1930s—industrialization by technology import in the private sector and creation of scientific manpower in the public sector. Technology import was encouraged by very liberal policies. Foreign capital and skill were needed and welcomed; no constraints on the nature of the industry or export obligations were imposed. Because of the small and isolated market, protection by import quotas and, later, by substantial tariffs was necessary. The result was the rapid establishment of an apparently balanced and highly diversified structure of import replacement industries in the 1930s and after World War II. The government encouraged local production by multinationals, particularly in the more technology-intensive areas of that era—transport, chemicals, petrochemicals, and early electronics—since local knowledge in these fields was not available.

The second, longer-term prong of government policy related to the creation of indigenous science. The small, new industries had neither the scale nor the resources to carry out their own R&D, nor could they have competed in volume or quality with the vast R&D establishments of multinationals that could supply technology on a marginal cost basis. The government therefore created science through the only means it could use to attain minimum viable scale—by concentrating scientific manpower in the universities and, even more so, in a large, steadily growing government establishment, the CSIRO,[5] and in the departments of agriculture and defense. This reflected the spirit of the time and the Oxbridge tradition that still dominated Australia—that science, left to itself, would automatically produce indigenous cutting-edge technology and industries. However, this process did not occur.

Essentially, two streams of science developed. The technology-intensive private sector became locked into the international network of the technology-generating companies, and government scientists were locked into the international publication race, largely in pure science. The smaller companies gained knowledge by diffusion and technical service from the larger organizations. The steel industry grew rapidly to become the dominant industry and established its own R&D. But the majority of the science-intensive industries were subsidiaries of multinationals, which provided an invaluable service to the country. Within two decades, from 1950 to 1970, they introduced modern manufacturing systems and became veritable universities of technology, marketing, and management.

This sequence of events was particularly true of the chemical industry, the one modern industry that was vertically integrated and had a sufficient concentration of resources to create substantial local research (25 percent of private sector R&D in manufacturing), development, and engineering organizations. Several companies established an effective system of joint overseas and local shareholding, local management with access to international skill, and interaction between this and local innovation. On balance, however, this mode of operation characterized the minority of companies. Many overseas companies then, as now, found it more efficient to operate in the branch plant mode—what the Canadians, in a similar situation, have termed the "truncated industry" mode—an industry with production and marketing but no local research, development, and engineering capacity, and so a much reduced capacity to innovate.

At the same time, between 70 and 80 percent of the best scientists were concentrated in the public sector. Their orientation to the centers of international science and strong scientific ethos produced much outstanding work, but most of it was in agriculture and mining, the community interest, and general science. Even now, no more than about a quarter of CSIRO's work is related to the new manufactures. The stronger the pure science ethos in the public sector became, the less able the private sector became to use its results, and the less interest it took. This process, in turn, reinforced the scientists' commitment to pure science.

The appropriate balance between scientific push and market pull is a contentious issue and varies between industries and countries. Economists have often made the point that fixed-factor technology transfer can be inappropriate if labor and capital costs differ between countries. Australia suffered from an inappropriate fixed-factor science structure, a higher ratio of basic science to applied R&D, and a higher ratio of scientists to engineers than most other nations. With 0.4 percent of the world's population, Australia has produced about 2 percent of the world's science—it is the eighth-largest producer of scientific papers—but owns only 0.3 percent of the world's intellectual property.

The belief in omnidirectional scientific progress remains strong. Only slowly is the recognition spreading that in small countries isolated scientific achievements do not transfer to local technology, but flow into the international pool of science. The stark contrast to this policy is, of course, the Japanese and South Korean concentration on application and engineering during the technology importation phase.

The problem of converting public-sector science into technology is encountered in most countries. In small countries, however, it is aggravated by the following considerations:

- Scale: a fragmented, diverse, import replacement-based manufacturing industry;
- The difficulty of competing with international skill, amortized on a world market basis;

- Liberal economic policies that negate targeting—"governments can't pick winners"; and
- Reluctance to target science by resource allocation.

Although major economic changes have occurred, both public- and private-sector science have been in the rear guard rather than the forefront of reorientation and reallocation of resources. The need for industrial and scientific reorientation was foreseen by some observers. But, it was the decline of trade of primary products that caused severe trade imbalances and a 40 percent devaluation of the Australian dollar (50 percent against the 1984 Japanese yen and the German mark) that jolted national awareness of the risks of technological obsolescence.

The problem is structural and so is broader than the science and technology issue. Nevertheless, national success or failure in converting science into a share of globalized technology is an important element. This conversion is particularly important at Australia's present level, between a resource-based and a more technology-based economy, burdened with a higher wage structure than those of the surrounding nations, which are approaching a similar transition. If the model of the product cycle is valid, the ability to innovate will determine whether the full industrialization level is attainable. To this extent, Australia's innovation problem has significance for other rapidly advancing nations in the Pacific and elsewhere.

Recent Measures

Economic pressures have recently produced important changes:

- Substantial government subsidies (150 percent tax deduction) for private-sector R&D and some R&D grants in energy and biotechnology;
- Government pressure on the private sector to raise R&D from 0.2 percent to more than 0.3 percent of the GDP in 1986 and on the public sector to reorient work toward the manufacturing industries;
- Change in CSIRO structure—management by a corporate-type board with a nonscientist chairman; and
- Generation of venture capital. The disadvantages of small scale can be overcome to some extent by creation of specialized niches based on local invention and individual entrepreneurs. Lack of venture capital and management skill are often the limiting factors. A Management Investment Company (M.I.C.) scheme, developed by the Australian Academy of Technological Sciences at the request of the Minister for Science, created a series of venture capital companies, half financed by government, which provide risk capital, management guidance, and marketing skills to new venture companies, most of which are R&D based. The M.I.C. scheme has stimulated inventors, entrepreneurs, and venture capital considerably.

Problems and Challenges

Australia has much to gain from further globalization: access to international technology, faster growth, improved competitiveness, and better use of scientific and engineering manpower. Her relative position in the region, however, is less certain. Rigidities in wage structure, labor organization, and both private and public administration may slow industrialization relative to that in neighboring nations.

Of most immediate practical importance—not necessarily in the context of innovation—is Australia's higher wage structure. Manufacturers in Australia, like those elsewhere, have transferred some of their production steps overseas in the interest of competitiveness of the overall process and harmonization of the international markets.

Australia, too, faces the worldwide problem of maintaining an adequate level of protection as a buffer against social upheaval, without, however, unduly slowing the growth of international trade. It is unlikely that in the near future most areas of manufacture will be internationally competitive, as mining and agriculture are now. The economy may therefore be able to bear a modest level of cross-subsidization in the form of residual tariffs for the labor-intensive industries to prevent permanent widespread unemployment. Similar dilemmas will be faced by other countries, including the United States, which are major producers of raw materials.

Other challenges include:

- The manufacturing industry needs to improve its international competitiveness and export performance, at least in some areas, to gain the advantage of scale.
- The private sector must increase its support of R&D as a means to achieve this goal.
- The public sector needs to increase its orientation to economic needs, particularly in the manufacturing sector.
- Local companies and subsidiaries need to step up cooperation in R&D with multinational companies.

The major driving force of globalization of industries is the multinational company. As the scientific potential of smaller and developing countries has become more important, their inclusion in the global process of technology creation has increased in importance. Their inclusion contributes already, but could contribute much more to maximizing creative potential worldwide. A practical and commercially acceptable approach is close interaction between technology donors and recipients in R&D. The underlying concept is plausible. Research is not a zero-sum game; both parties must gain. Know-how is tradable, and improved personal communication is making know-how trading much easier. Synergism between donor and recipient presupposes the following conditions:

- Benefit to both—for example, proximity to customers for the licensor and ability to build on international know-how for the licensee;
- Sufficient scale and competence of licensee to make synergy likely; and
- A sound patent system and respect for confidentiality.

Several British companies, perhaps because they have long been associated with Australian companies, have evolved such patterns of collaboration in R&D with their subsidiaries in Australia. The arrangements provide for free exchange of information, but the right of use, including export, is negotiated case by case on the basis of optimum fit. In selected areas, in which the local subsidiary had concentrated or made a major invention, say in veterinary drugs, electrical tools, or mining technology, the subsidiary is given a corporate mandate, global or regional. Overall, however, progress in R&D collaboration has been disappointingly small, partly because of inexperience but partly also because of bureaucratic rigidities in corporations and government.

UNITED STATES AND JAPAN: ENGINES OF GROWTH IN THE PACIFIC

Investment and Technology Transfer Strategies

On the national evolutionary scale of the production cycle, the United States and Japan occupy the top positions. Both nations are passing through the transition from an economy based on mass production to one based on information technology and its application to mass and customized manufacture.[6] Their capital (see Figure 4) and innovation provide the push- and pull-through of industrialization in the Pacific region. Nevertheless, as a proportion of the U.S. worldwide investment, investment in the developing nations of the Pacific, excluding South America and the People's Republic of China, is small. The U.S. royalty income from licenses of U.S. affiliates is also modest, about $259 million, or 5 percent of the $5,042 million total in 1979 (Hill and Johns, 1983).

There are differences in investment and technology transfer between the two countries. One notable difference is the substantial Japanese investment in the United States, in contrast to the small U.S. investment in Japan. Japanese investment is stronger in ASEAN, and American investment predominates in Oceania and Canada. These patterns may be explained in relation to geography, economic structure, and historical links. There are also differences in technology export. U.S. exports are concentrated in the more technology- and capital-intensive areas, whereas Japanese technology exports

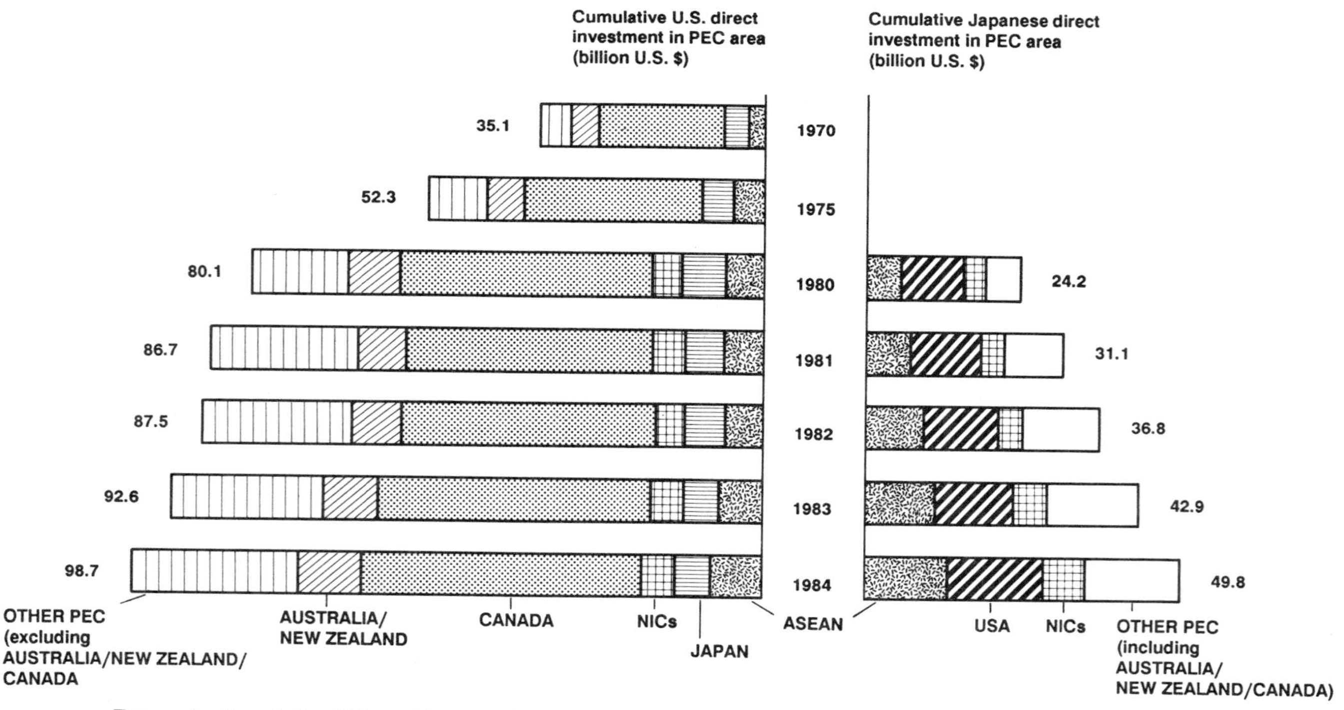

FIGURE 4 Cumulative U.S. and Japanese investment in Pacific Economic Community area. SOURCE: Buchanan (1986).

focus on the less developed nations and, appropriately, the more labor-intensive activities. This disparity largely reflects the different stages of technological development of the two countries—particularly the greater differences of the past. It is probably also the result of greater Japanese reluctance to transfer high-technology knowledge. In contrast to the United States, which has long been the world's largest net exporter of technology, Japan's express policy of importing foreign technology has made that nation a net importer of technology for some decades, and these imports are still growing. However, the crossover point has been reached; in payments and receipts for new licenses, Japan became a net exporter in the mid-1970s.

Both the United States and Japan have transferred some of their labor-intensive industries to less developed nations. The United States was guided in this action primarily by economic considerations. Transfers included important electronics production steps, which were moved to the NICs in the Pacific. Japan has criticized this policy (Saba, 1986) and has mainly transferred more mature processes.

Sources of Competition

Some of the elements of cooperation and competition between the United States and Japan have been highlighted in recent trade controversies. The two nations hold some common beliefs: internationalization of industries, deregulation, free trade, fair competition, and concentration on the knowledge-based industries as the economic driving force. Nevertheless, there have been tensions between the two countries, with the current confrontation centered on semiconductors, the key to informatics, which have a tremendous impact on manufacturing and trade.

The respective bases of competition are different. The United States unquestionably pioneered semiconductors and still leads in the science base. The United States grants 10 times as many Ph.D.'s in science and engineering as Japan, and Japanese university-based R&D expenditures are barely one-fourth those of the United States (Saxonhouse, 1986). Thousands of Japanese students study in the United States, and research projects financed by Japanese companies in U.S. universities are important sources of information. The United States also leads by a strong margin in aerospace, software, computing, pharmaceuticals, and medical equipment.

The Japanese have gained advances over the United States in production technology, color televisions, video cassette recorders, cars, and—the current issue—in mass production micron- and submicron-sized microlithography. Even in mature industries where the United States and Western Europeans innovate by incorporating microprocessors, the Japanese have invented new systems approaches—greater attention to quality, timing, and customizing products by using the flexibility of very-large-scale integration in production processes.

In some ways, the U.S. emphasis on science and the Japanese preoccupation with application are reminiscent of earlier positions of the United Kingdom and the United States. They reflect the research structure of the past, particularly Japan's almost total focus on development (54.0 percent of current R&D expenditures) and applied research (43.7 percent) and little focus on basic research (2.3 percent).

However, the situation is changing. As Japan has drawn level with Western science, the "fastest follower" policy, that is, government-supported programs introducing Western technology, has become less effective. As stated by the Japanese Ministry of International Trade and Industry (1980), "In the past, Japanese industry achieved brilliant results in improving and applying imported technologies. In the 1980s, however, it will be essential for Japan to develop technologies of its own." The first of three MITI policies will be the development of creative technologies, involving an increase in basic R&D to 12.5 percent of total R&D funds.

Both nations have taken confrontationist positions in regard to some industrial practices. U.S. producers have attacked targeting, calling it "coordinated government action that directs productive sources to give domestic producers in selected nonagricultural industries a competitive advantage" (Noyce and Wolff, 1986). This advantage, they claim, has led to deliberate overproduction in Japan and consequent dumping overseas. The Japanese have stated that targeting is no longer practiced and that "administrative guidance" is now no more than industry and government agreement to basic solutions. On the other hand, in the United States, government R&D contracts in defense and space and a large and entrepreneurial venture capital market produce a more than comparable concentration of resources.

Voices of compromise and cooperation have emerged. Toshiba's former chairman, Shoichi Saba, has suggested greater technology exchange and claims to have achieved collaboration with U.S. and Western European producers—Intel, AT&T, ITT, Siemens, and Olivetti (Saba, 1986). Car manufacturers have coordinated and rationalized production internationally. MITI, in its "Vision of 1980's Policies," has defined international cooperation in technological development as a major objective not only in the relatively noncompetitive fields of energy and food, but also in large-scale projects such as aircraft and marine exploitation, admittedly areas of U.S. rather than Japanese strength (MITI, 1980). Some see this as a Japanese two-track policy to expand into niches in the frontiers autonomously and to strengthen the Japan–U.S. collaborative system in areas of U.S. strength (Inoguchi, 1986).

Japan's current surplus in foreign currency, with a phenomenal U.S. trade deficit of $166 billion in 1986, has led to calls by her trading partners and the Japanese themselves for greater domestic consumption. This increase in consumption is possible, but with current prosperity, Japanese households seem to be as cluttered with manufactured goods as those of Japan's Western competitors. Housing is the one area in which Japanese living standards appear to be inferior and where there is scope for much improvement—larger and better apartments and individual houses are needed. This deficiency seems to have received remarkably little attention, perhaps because the mass-manufactured component of it is relatively small.

Japanese predictions of future development differ little from those of the Western world. The present phase of "mechatronics"—the integration of machines and electronics (Makino, 1984)—is expected to lead to a $1-trillion world market in electronics, telecommunications, and computers in the 1990s, of which Japan's share is expected to be in the $100-billion range (Noyce and Wolff, 1986).

Perhaps the ultimate field of international competition and cooperation will be education. Japan has almost 900 colleges and universities, more than any other nation but the United States and more per capita than any other nation. Now, as in the past, Japan has placed far greater emphasis on engineering than science. Research has been primarily done by private industry, more so than in any other Western nation except, perhaps, Switzerland. It has therefore been product effective rather than spectacularly innovative. However, Japanese dedication to education is unsurpassed, and the impact of education policy changes in the past has been sharp and fast. The policy of more education and more basic research may well produce spectacular results in the future.

Problems and Challenges

Both countries, particularly the United States, need to pursue innovative technology to facilitate Pacific regional industrialization. Although U.S. investment in the region has been valuable, it is only a small portion of its global investment and has produced remarkable results. Hence, one would hope that this investment will be maintained, despite the fact that it has indirectly induced competition from Japan and South Korea. Many view the continued drive of the U.S. and Japanese economic dynamos and resulting growth and prosperity as the most important insurance against political instability. Finally, many observers also see the increasing inclusion of the scientific and technological manpower of the less developed nations of the region in the global process of technological creativity spearheaded by Japan and the United States as an important next step in development.

CONCLUSION

Globalization of industries by technology flow to the Pacific region has proceeded faster than in other regions if measured by the GDP growth it induced. It has been facilitated by differing levels of development, comparative advantage, and endowment of raw materials among the countries of the region. Rates of development have also been fast because base levels in some countries are still low and perhaps even more so because many of the economies are open and native skill has been able to cope with rapid change. Indeed, national characteristics such as the Japanese focus on high-quality labor have modified industrial concepts.

The process of industrialization in many areas is far from finished. As countries move to the next plateau of development, "crowding" will become more severe and ability to innovate will be more important. Hence, achieving national ambitions will call for greater shares in international technology. Finally, the People's Republic of China's eventual industrialization will have an immense effect on its neighbors, but at present, the extent and timing of these developments defy prediction.

Successful industrialization in the region is particularly interesting, because it was faced with many impediments: large differences in development, language, culture, and isolation. Yet the problems of control of technology transfer have not evoked heated policy responses as they have in Latin America, Africa, and even some Western European countries, such as France in the 1950s and 1960s. In this sense, the report from the Pacific is, perhaps, an encouragement to other regions and an indication that we are getting better at technology transfer—competition coexisting with cooperation.

ACKNOWLEDGMENT

The most effective and willing help of Miss Corinne Boyles, of the Australia-Japan Research Centre, is gratefully acknowledged. Miss Boyles selected and assembled much of the statistical data and translated the Japanese tables used.

I would like to thank Dr. Peter Drysdale, director of the Australia-Japan Research Centre, Research School of Pacific Studies, Australian National University, Canberra, who advised on the selection of literature and made the resources of his center available.

I am also indebted to a number of scholars who provided literature and statistical data, in particular Mr. Clive Hughes, Economic and Social Commission for Asia and the Pacific, United Nations, Bangkok; Professor B. L. Johns, Director, Bureau of Industry Economics, Canberra; Dr. D. A. Kelly, Visitor to the Contemporary China Centre, Australian National University; Dr. D. McEwan, Department of Science, Canberra; Dr. Jorge L. Reys, National Science and Technology Authority, Manila; Dr. Sanga Sahasri, Ministry of Science, Technology and Energy, Bangkok; and Mr. Chihiro Watanabe, Chief Representative of the New Development Organization (Japan) in Australia, Sydney.

I also wish to thank Ms. Janet Muroyama, who edited the paper, and Mrs. Teresa Twine, who typed it.

NOTES

1. The definition of Pacific Economic Community (PEC) used here follows the tables issued by the Pacific Basin Economic Community Member Committee of the Tokyo Chamber of Commerce and Industry International Division, May 1986. Elsewhere, this paper focuses on representative countries of the Pacific Rim—the Association of Southeast Asian Nations (ASEAN)—Indonesia, Malaysia, Philippines, Singapore, Thailand, and Brunei, the newly industrializing Asian nations (Hong Kong, Singapore, South Korea, and Taiwan), Australia, Japan, New Zealand, and the United States. The major developed countries are dealt with only in relation to the region; Latin American countries are excluded from the paper because they are discussed elsewhere in this volume. The People's Republic of China is excluded, as its present impact on the region in proportion to its size and future role is small and as it is a centrally planned economy. Clearly, it represents a topic of its own and is beyond the scope of the present discussion. Because the data available differ in definitions of the region, groupings in tables are not always consistent.
2. A more sophisticated but still simplified picture of the product cycle and trade interaction applied to the People's Republic of China–Pacific Economic Region has been presented by Findlay et al. (1986).
3. Indonesia, Malaysia, the Philippines, Singapore, and Thailand. Brunei, the sixth member of ASEAN, is excluded from these considerations because of its small size and unique character. It is best described as an emirate-type oil economy of 200,000 inhabitants with a per capita income of $24,000.
4. For a complete statistical analysis of the potentially negative impact of direct foreign investment, see Park (1984).
5. The Commonwealth Scientific and Industrial Research Organization represents 15 percent of Australia's total national R&D. By comparison, the Max Planck and Fraunhofer Institutes together represent 2.5 percent of total R&D in the Federal Republic of Germany.
6. The exclusive emphasis on the "information society" is somewhat futuristic, because much of the so-called information revolution still revolves around manufacture—chips and computers—and their incorporation and servicing in manufacture and trade.

REFERENCES

Buchanan, I. C. 1986. The ASEAN Experience. Paper presented at the Stanford Research Institute International Associates Program, Helsinki, 8–10 September 1986.

Drucker, P. 1986. The changed world economy. Foreign Affairs 4(4):768–791.

Earnshaw, P. 1980. The Identification of Lace. Aylesbury, United Kingdom: Shire Publications.

Economic and Social Commission for Asia and the Pacific (ESCAP). 1984. Technology for Development. Study by ESCAP Secretariat for the Fortieth Session of ESCAP, 17–27 April 1984. Tokyo: ESCAP.

Findlay, C., K. Anderson, and P. Drysdale. 1986. China's Trade and Pacific Economic Growth. Paper prepared for the Australia-Japan Research Centre Workshop on Australia, China, and Japan, Australian National University, Canberra, 19 September 1986.

Harbison, F., and C. A. Myers. 1964. Education, Manpower and Economic Growth. New York: McGraw-Hill.

Helleiner, G. D. 1979. Transnational corporations and trade structure. University of Toronto, Toronto, Ontario, Canada. Memorandum.
Hill, H., and B. Johns. 1983. The transfer of industrial technology to western Pacific developing countries. Prometheus 1:1.
Hufbauer, G. C. 1970. The impact of national characteristics and technology on the commodity composition of trade in manufactured goods. In The Technology Factor in International Trade, R. Vernon, ed. New York: Columbia University Press.
Inoguchi, T. 1986. Technology Security and Trade: An Emerging Nexus in East Asia. Canberra, Australia: Institute of Oriental Culture, University of Tokyo and Australia-Japan Research Centre, Australian National University.
Kim, L. 1984. Technology Transfer and R&D in Korea: National Policies and the U.S.-Korea Link. Paper presented at the Conference for National Policies for Technology Transfer, Hawaii, October 8.
Kuznets, P. W. 1984. Economic development, export structure and shifting comparative advantage in the Pacific region. Pp. 42–51 in The Industrial Future of the Pacific Basin, R. Benjamin and R. T. Kudrle, eds. Boulder, Colo.: Westview Press.
Makino, N. 1984. High technology in Japan: Its present and future. Nippon Steel Forum April:14.
Ministry of International Trade and Industry. 1980. The Vision of MITI Policies in the 1980's. Tokyo: MITI Information Office.
Noyce, R. N., and A. W. Wolff. 1986. High-tech trade in the 1980s: The international challenge and the U.S. response. Issues in Science and Technology 2(Spring):61–71.
Onishi, A., and O. Nakamura. 1986. Long-term economic perspectives for Asia and the Pacific 1986–2000. Center for Global Modelling, Joka University, Tokyo.
Park, E. Y. 1984. Patterns of foreign direct investment, foreign ownership, and industrial performance: The case of the Korean manufacturing industry. Pp. 129–134 in The Industrial Future of the Pacific Basin, R. Benjamin and R. T. Kudrle, eds. Boulder, Colo.: Westview Press.
Reynolds, L. G. 1985. Economic Growth in the Third World 1850–1980. New Haven, Conn.: Yale University Press.
Saba, S. 1986. The U.S. and Japanese electronics industries: Competition and cooperation. Issues in Science and Technology 2(Spring):53–60.
Saxonhouse, G. R. 1986. Why Japan is winning. Issues in Science and Technology 2(Spring): 72–80.
Sekiguchi, S., and T. Horiuchi. 1984. Foreign trade and industrial policies: A review of Japanese experience. In the Industrial Future of the Pacific Basin, R. Benjamin and R. T. Kudrle, eds. Boulder, Colo.: Westview Press.
Vernon, R. 1966. International investment and international trade in the product cycle. Quarterly Journal of Economics 80(May):190–200.
World Bank. 1984. World Bank Development Report 1984. Washington, D.C.: World Bank.
World Bank. 1985. World Bank Development Report 1985. Washington, D.C.: World Bank.
World Bank. 1986. World Bank Development Report 1986. Washington, D.C.: World Bank.

Technology and the World Economy: The Case of the American Hemisphere

ENRIQUE MARTÍN DEL CAMPO

WE LIVE IN A COMPLEX INTERNATIONAL setting in which technology has an ever-growing influence on the world economy. The diffusion of technologies is one factor in the globalization of the marketplace which influences the transculturation process, resulting in shifts in national preferences, values, and aspirations. The progressive pluralism in international trade contrasts sharply with the growing bipolarity in the strategic and military affairs of the two superpowers, the United States and the Soviet Union. From a strictly economic standpoint, the heightened struggle between the two superpowers diverts resources that both countries could more usefully apply to the problems of their relationships with other countries.

The development of the Latin American and Caribbean countries is affected by these continuously changing international influences. Many of the problems confronting our societies are linked to such international phenomena as the debt crisis, which affects both debtors and creditors. Despite these problems, Latin American and Caribbean countries are engaging in greater international activity, and their policies and proposals are receiving more attention in hemispheric and global affairs.

The region is heterogeneous and includes countries of varied complexity and history. Despite these differences, a basic solidarity is evident in their international actions, and the region is perceived by the rest of the world as

This paper was written on the basis of a draft by Zoltán Szabo. Also assisting with the intermediate documentation and the final version were Raúl Allard, Arturo Garzón, Washington Monge, and Rubén Perina. The authors are staff members of the Organization of American States, and none of the views expressed in this study reflects explicitly the positions of the General Secretariat of the organization.

a unified whole. This gives Latin America a particular capacity for action and negotiation that can be best used by setting joint and independent positions.

LATIN AMERICA AND THE CARIBBEAN IN THE WORLD ECONOMY

A notable feature in the development of the world economy is the persistent inequality of the Latin American and Caribbean countries' economic weight. The gross domestic product of the United States, the Soviet Union, or Japan is 1,000 times larger than those of the small countries of the world, and up to 10,000 times larger than those of the smallest national economies. The countries of Latin America and the Caribbean, except for Brazil and Mexico, are in the group that is 100, 1,000, or up to 10,000 times smaller than the superpowers (Table 1).

The economic weight of all developing countries, Latin American and Caribbean countries in particular, has hardly increased at all in the last quarter century. From 1960 to 1982 Latin America's share of world production rose from 4.2 to 5.3 percent (Table 2). This increase was mainly due to Brazil and Mexico, whose share of world production rose from 1.1 to 1.8 percent and from 0.9 to 1.3 percent, respectively.[1] Most of the other countries in the region experienced only slight increases or declines in their share of world production.

The relative weights of the market economy in developed countries and in countries with centrally planned economies have remained virtually unchanged since 1960. However, changes within each group have significantly affected the international economic and political situation. At the same time that the U.S. share dropped between 1960 and 1982 from 36 to 26 percent, Japan's share rose from 3 to 9 percent (Table 3). The change was even more dramatic in their relative power positions in the international economy. For 1985, Japan's exports totaled $174 billion and U.S. exports totaled $217 billion. But Japan showed a net foreign investment of $125 billion, whereas the United States recorded a net foreign debt of $107 billion (Brown et al., 1986; World Bank, 1986). Consequently, Japan's external economic position, measured only with these partial indicators, appears better than that of the United States.

Japan's economic power is supported by the combination of a high propensity to save and invest domestically, low military expenditures (1 percent or less of gross domestic product), and an orientation toward building a large

[1]For further information see *Yearbook of National Accounts Statistics for 1976, Volume II, International Tables,* United Nations, New York, 1976; *National Accounts Statistics: Analysis of Main Aggregate, 1982,* United Nations, New York, 1985; *Statistical Bulletin,* January–December 1985, Organization of American States, Washington, D.C.; and Brown et al. (1986).

TABLE 1 Economic Magnitude of 120 Countries Measured by Gross National Product, 1982 (in U.S. dollars)

More than $1,000 billion
1. United States
2. Soviet Union
3. Japan

$200 billion to $1,000 billion
4. Federal Republic of Germany
5. France
6. United Kingdom
7. People's Republic of China
8. Italy
9. Canada
10. Brazil

$100 billion to $200 billion
11. Spain
12. India
13. Australia
14. Mexico
15. Democratic Republic of Germany
16. The Netherlands
17. Saudi Arabia
18. Romania
19. Poland

$50 billion to $100 billion
20. Sweden
21. Switzerland
22. Indonesia
23. Belgium
24. Yugoslavia
25. South Africa
26. Nigeria
27. Czechoslovakia
28. Republic of Korea
29. Austria
30. Denmark
31. Norway
32. Argentina
33. Turkey

$10 billion to $50 billion
34. Finland
35. Bulgaria
36. Algeria
37. Venezuela
38. Philippines
39. Greece
40. Thailand
41. Hungary
42. Egypt

$10 billion to $50 billion (*continued*)
43. United Arab Emirates
44. Pakistan
45. Colombia
46. Hong Kong
47. Malaysia
48. Cuba
49. Israel
50. New Zealand
51. Peru
52. Kuwait
53. Chile
54. Syria
55. Republic of Ireland
56. Singapore
57. Morocco
58. Bangladesh
59. Ecuador

$1 billion to $10 billion
60. Guatemala
61. Ghana
62. Tunisia
63. Qatar
64. Ivory Coast
65. Oman
66. Dominican Republic
67. Sudan
68. Zimbabwe
69. Uruguay
70. Kenya
71. Burma
72. Paraguay
73. Republic of Zaire
74. Tanzania
75. Sri Lanka
76. Ethiopia
77. Panama
78. Jordan
79. Brunei
80. Jamaica
81. Gabon
82. Costa Rica
83. Zambia
84. Trinidad and Tobago
85. Luxemburg
86. El Salvador
87. Bolivia
88. Madagascar

(*continued*)

TABLE 1 (*Continued*)

$1 billion to $10 billion (*continued*)	Less than $1 billion
89. Honduras	107. New Caledonia
90. Nicaragua	108. Liberia
91. Iceland	109. Togo
92. Senegal	110. Barbados
93. Papua New Guinea	111. Swaziland
94. Nepal	112. Guyana
95. Cyprus	113. Lesotho
96. Reunion	114. Belize
97. Haiti	115. Seychelles
98. Bahamas	116. Antigua and Barbuda
99. Rwanda	117. Grenada
100. Malawi	118. St. Vincent and the Grenadines
101. Sierra Leone	119. Dominica
102. Fiji	120. Tonga
103. Malta	
104. Burkina Faso	
105. Mauritius	
106. Burundi	

SOURCES: Organization of American States and the World Bank.

technological and industrial potential supported on a firm base of education, science, and technology. The dramatic expansion of Japan and the significant contraction of the U.S. share have been accompanied by an increase in the Soviet Union's economic share in the countries with centrally planned economies.

In this context, there is economic, technological, and ideological competition between the developed market economies and the centrally planned economies. At the same time, the first group experiences trade wars and struggles that create difficulties for the developing countries, particularly when these events lead to trade and technology protection measures that produce unfavorable shifts in international trade.

TABLE 2 Breakdown of World Gross Domestic Product by Groups of Countries, 1960 and 1982

	Percentage of World Gross Domestic Product in:	
Countries	1960	1982
Developed countries	65.8	64.8
Centrally planned economies	19.8	19.0
Developing countries	10.2	10.8
Latin America	4.2	5.3

SOURCES: Organization of American States and the World Bank.

TABLE 3 Contribution of the 10 Largest National Economies to the World Gross Domestic Product

Country	Percent	
	1960	1982
United States	36.0	25.6
USSR	7.7	10.3
Japan	3.1	8.9
Federal Republic of Germany	5.1	5.5
France	4.3	4.5
United Kingdom	5.1	4.0
People's Republic of China	6.8	3.2
Italy	2.5	2.9
Canada	2.8	2.5
Brazil	1.1	1.8
All other countries	25.6	30.8

SOURCES: Organization of American States and the World Bank.

The present international situation, then, presages change for the future. Increasingly, the developing countries are demanding a fairer and more equitable share in the international economy. Given this trend, Latin American and Caribbean countries will need to learn how to benefit from emerging opportunities and overcome new problems. Also desirable is a dual strategy for outward growth, with optimum participation in the international economy and, within the region, through subregional and national markets. Both inward and outward growth will require improved technological capacity, entrepreneurial activity, and rational government participation.

HIGHLIGHTS OF WORLD TECHNOLOGICAL DEVELOPMENT

Research and Development Expenditures

Closely connected with the changes in the world economic weight of the larger industrialized countries is a change in their technological and industrial weight. Japan and the Federal Republic of Germany now exceed the United States in their rates of research and development (R&D) expenditures devoted directly to technologies with industrial applications. They achieved this by shrinking expenditures on military R&D. As a result, both countries have substantially increased their technological and industrial potential and have become highly effective competitors of the United States (Table 4).

The rate of expenditures on research and development in industrialized countries has an impact on developing nations as well. The higher rate of technological innovation in developed countries makes it possible for them to maintain a privileged position in international markets, particularly with respect to levels of demand and profitability.

TABLE 4 Percentage Share of Military Programs in the Government Research and Development Budget of Selected Countries

Country	1961	1976
United States	71	50
United Kingdom	65	48
France	44	30
Federal Republic of Germany	22	11
Japan	4	2

SOURCE: Norman (1981).

The United States accounts for approximately one-third of the world's expenditures on R&D, or somewhat over $150 billion a year. Western Europe and Japan together account for another third, and the Soviet Union and the centrally planned market economy countries in Eastern Europe account for most of the rest. According to a recent estimate, the developing countries account for less than 3 percent of the total.[2]

By the start of the 1980s, research and development expenditures in Latin America and the Caribbean were estimated at about $1.5 billion, or 1 percent of the world's total (Sagasti and Cook, 1985). Thus, the ratio of research and development expenditures to gross domestic product was only 0.24 percent in the region at the start of the 1980s, compared with 1.8 to 2.3 percent for most of the larger industrialized countries. The per capita cost of R&D in these countries was $200, whereas the average for Latin America was less than $5 (Norman, 1981).

About 3 million scientists and engineers are employed in research and development worldwide. In 1980, Latin America had about 100,000 researchers, or 3.3 percent of the world total. Thus, the region had less than 350 researchers per million inhabitants, compared with a world average of 673. The Latin American average was less than one-third of the U.S. figure (Martín del Campo, 1983).

Research and Development and Comparative Advantage in International Trade

Accelerated worldwide technical change, based on intensive research and development in the industrialized countries, is modifying the conditions of comparative advantage in international trade. Two of these modifications are particularly important. First, comparative advantage is no longer defined

[2]For further information see Martín del Campo (1983), Norman (1981), and Sagasti and Cook (1985).

only by the abundance and relative cost of the traditional factors of labor, capital, and certain natural resources; it is also defined by the technological capacity to produce and sell new or diversified products. Second, the development of science-based technologies is completely revolutionizing modes of production. Microelectronics and informatics, biotechnology, and the production of materials with special properties, among others, are changing the relative use of capital and labor in various productive sectors. González (1986) has pointed out that Latin America's comparative advantage in labor-intensive production and natural resources is being eroded by the introduction of new technologies in the developed countries. Hence, it is essential to understand such technological transformations and to define the actions needed to support appropriate domestic activities that respond to the new situation and develop a more diversified export structure.

The region's technology strategy cannot concentrate on a few advanced sectors and a few productive strata in the sectors since such concentration would be inconsistent with the objective of improving economic and social conditions of broad sectors of the population. It would be more beneficial to promote and develop technology in traditional small- and medium-size enterprises, both rural and urban, by adapting technologies that are purchased abroad and by adjusting to intermediate technologies. Such technologies include those that emphasize manpower rather than modern concepts, e.g., factory drill presses without numerical control.

These more innovative strategies have three required elements. First, the development and commercial application of more advanced key technologies—which require a greater concentration of resources—must be accompanied by modest innovations with more immediate applications. Second, a high proportion of small and medium enterprises should be linked to large enterprises through subcontracting relationships. Third, technological development should not be concentrated in a few sectors since that dooms the other sectors to obsolescence in a free international market.

In practice, it is appropriate for the region to choose a combination of various types of technologies. These would include certain state-of-the-art technologies (microelectronics, biotechnology, and new materials) and, at the other extreme, technologies to increase the use of labor and support marginal sectors. The mix of technologies would be determined by the relative development level of each country and its objectives. These issues will be explored in greater detail in a later section of this paper.

PARTICIPATION OF THE REGION'S COUNTRIES IN THE WORLD ECONOMY

The economies of the Latin American and Caribbean countries participate in the world economy mainly through foreign trade. Other economic growth variables such as investment, domestic demand, production, and technology are determined by the extent of participation in foreign trade.

Roots of Asymmetrical Interdependence

After the postcolonial period, the countries of the region took part in the world economy according to an international division of labor based on their comparative advantage in producing and exporting agricultural and mineral raw materials and importing manufactured goods. When international prices were high for these raw materials, the resulting prosperity promoted the creation of the infrastructure needed to exploit, transport, and export the goods and produced development in many other sectors.

However, the developed countries displayed a long-term trend toward consuming manufactured products, which accounted for most of the region's imports, and slow growth in the demand for primary-sector products, which represented the overwhelming proportion of their exports. This resulted in firm prices for industrial products and a relative reduction in the prices of raw materials. Thus, the countries were faced with a chronic terms-of-trade problem.

As the region lost the advantages of the existing international division of labor, trade became more asymmetrical and development more subject to changes in the markets of industrialized countries. While those countries took the lead and assumed a growing weight in the world economy, the countries of Latin America lagged behind.

By 1940, the developed countries were establishing and maintaining decisive advantages and influences over Latin American and Caribbean countries in technology, education, culture, politics, and defense. Consequently, the asymmetry went beyond commerce to include all of the interactions between the region and the developed countries. It is important to keep in mind that this phenomenon is one of asymmetrical interdependence rather than direct dependence on the developed countries, for the pace of development in the "center" was determined largely by their economic ties with the peripheral countries.

Now, increasing integration of the world economy, involving trade and financial and geopolitical relations, has given peripheral countries greater potential for maneuvering, negotiating, and making independent decisions in their relations with the developed countries. This particularly favors those countries that have already achieved high intermediate development, a category that includes most of the countries in the region.

Development Patterns: 1940–1975

In the 1940s and 1950s, when the limitations of preferential production and export of a few raw materials became clear, the Latin American and

Caribbean countries adopted an industrialization policy based on import substitution and product diversification, a strategy of inward growth. Thus, manufacturing production increased in the region at an annual rate of slightly over 6 percent in the 1950s (Schwartz, 1983).

This growth followed a development pattern that mimicked the consumer society models of the developed countries. There was a clear preference also for developing lines of products with markets in the wealthy sectors of the population and neglecting to meet the massive needs of the poorer sectors for products adequate in type, cost, standard, and quality. This growth pattern was partly shaped by certain regressive economic, political, and social structures and also by transnational corporate interests from the developed countries. Manufactures increased by 6.5 percent between 1960 and 1965.

In the second half of the 1960s, industrialization in the region began to enter a more mature phase and started a new tentative phase of "growth outward." Manufactures grew at a rate of 7.5 percent in that period. The region saw the beginning of a policy of developing nontraditional exports, particularly manufactured goods. From 1966 to the mid-1970s, the share of manufactured goods in total exports increased 10-fold, rising from 2 to 20 percent, excluding processed foods and metals, and generally production grew by about 8.5 percent. If these items are included (most should strictly be classified as semimanufactured goods), the share of industrially processed exports rises to nearly 55 percent. By 1975 manufacturing production in the region began to slow, and in the early 1980s there was a clear recession (Schwartz, 1983).

In the outward growth phase, the asymmetry of the region's trade in terms of global technological innovation underwent a transition. Latin America began to export products with a relatively low degree of technological sophistication, although they were not necessarily raw materials, whereas its imports from advanced countries had a larger content of new technology. Thus, the asymmetry of trade took a new form, still detrimental to the region.

Reversing Asymmetry in Trade

There are a number of significant factors which affect the competitiveness of the countries of the Latin American and Caribbean region in foreign markets. These include rate of innovation, the ability to apply advanced technology, degree of capital investment, use of natural resources, and the existence of technological support services.

National producers must be able to penetrate foreign markets with new and diverse products, which must be produced and delivered at low cost, with satisfactory quality, and on time. To accomplish these things, companies require favorable conditions of national economic, industrial, and technological policy. They must also overcome their management problems. The more industrially advanced countries of the region must also try to overcome, slowly at first, their current inability to originate innovative products that incorporate advanced technologies. These products are desirable because their export is not very vulnerable to the troughs of the economic cycle. To promote technological innovation to achieve these goals, creativity and innovative change must assume cultural value among scientists, businessmen, and technicians, as well as factory and farm workers. This is perhaps the most complex challenge, and it requires many changes in the education system at all levels, from primary school to higher education (Waissbluth, 1983).

A second factor affecting the region's competitiveness in world markets is its ability to apply advanced technologies. The region's comparative advantage in labor-intensive production processes that use local natural resources is being eroded by state-of-the-art advanced technologies, which are widely used in developed countries. The impact of technologies such as microelectronics, informatics, robotics, biotechnology, and recently developed special materials is particularly important since these technologies can be used to improve production and design of a wide range of products. Some countries in the region have the technical and financial capacity to adapt these applications to the local environment or to develop new applications. They would benefit from participating with advanced countries in developing applications of state-of-the-art technologies.

Moreover, these countries must develop a strong domestic scientific base in order to best utilize these advanced technologies. The boundaries between science and technology are disappearing as marketable technologies and scientific know-how take their place alongside industrial products as tradable commodities. Consequently, it is becoming essential that the workers who deal with these technologies in a production environment be trained in basic as well as applied science.

The rate of capital investment in the region is a third factor affecting Latin American and Caribbean countries' ability to use technology and develop productive export sectors. Most of the machinery and equipment used by companies in the region are worn out or outdated, and solving this problem would be costly. Between 1986 and 1995, the region would need about $600 billion for new machinery and equipment to modernize and expand its productive capacity and achieve an average annual growth of 4 to 5 percent in gross domestic product. Given this high cost, it is necessary to increase domestic design and production capacity for machinery in industries using the entire range of technologies: labor-intensive, intermediate, and advanced.

At the same time, the countries must meet the terms for foreign purchase of capital goods, arrange for the transfer of technology connected with those goods, and implement mechanisms for intraregional cooperation and trade. Argentina, Brazil, and Mexico have already undertaken cooperative efforts in this field with LATINEQUIP, a technological, industrial, and financial consortium. Argentina, Brazil, and Uruguay also have recently begun an ambitious economic integration experiment.

A fourth factor giving impetus to exports and economic growth is that the Latin American and Caribbean countries have a wide range of raw materials, whose transformation into processed products using recent scientific advances could be a nucleus of new technological and industrial development. This is particularly important for the small countries, which need open economies and specialization even more so than do the larger and middle-size countries in the region. The specializations of small countries in the Caribbean and Central America, for example, could be based on export-oriented production complexes using sugar cane and other food-oriented agro-industries. Such complexes could achieve a high level of international competitiveness by comprehensive use of basic raw materials.

Finally, technological support services are also necessary if a modern production system is to meet the requirements of foreign competitiveness in overseas and domestic markets. These include a technical information services infrastructure; research and development to replace products and processes, as well as develop new products; and preparation for—or at least the application of—basic engineering, quality control, product analysis, and standardization. Thus, to modernize and adapt to foreign competition, companies of all sizes need well-equipped laboratories and other national or local scientific and technological services to support them.

DOMESTIC DEMAND, PRODUCTION, AND TECHNOLOGY

Domestic demand in Latin America is limited by persistent internal economic and social problems such as obsolete land ownership systems, a markedly unequal income distribution, inequities of ethnic integration, segmentation of production and marketing systems, inequities for various socioeconomic groups in education and training, and restricted social mobility. Because these situations have been difficult to change, it has not been possible to filter the fruits of economic growth to all sectors of Latin American societies. In some cases, policy formulations have been designed to orient economic growth toward comprehensive social improvement, increased income for the poor, and the consequent expansion of domestic demand for a wide range of consumer products. However, these efforts have been hampered by social conditions and serious political obstacles and pressures.

Economic and technological variables also influence development in the countries of the region. An interpretation of these variables by Raul Prebisch and collaborators has recently been summarized by Marí (1986). In essence, technology can have both positive and negative effects.

On the positive side, technical changes in production systems tend to increase productivity and extend the production processes, leading to end products that are increasingly refined. The longer processes use increasing volumes of capital goods and materials, whose production involves increasingly complex and sophisticated technology.

A direct effect of increases in productivity is the decline in human labor requirements. In the developed countries, labor displaced in producing final consumer goods is reabsorbed in producing the new advanced production equipment. However, this does not occur in Latin American and Caribbean countries, which produce few capital goods or critical basic intermediary goods. Since the countries of the region must import such means of production, the technical change produces unemployment that is not reabsorbed.

The continued structural unemployment in the countries of the region perpetuates low wages and a small domestic market. In addition, if the demand for simple consumer goods is reduced because of low wages, the domestic market is biased toward consumption by the middle and upper classes. These are two aspects of unequal income distribution.

This biased composition of domestic demand leads to excess production (and imports) of consumer durables and shifts the stimuli to sectors with more imported technology and profitability, such as the sectors that produce means of production. The bias also restricts the development of those industries that would more easily produce domestically the necessary intermediate goods, which are simpler to produce than sophisticated consumer goods. These industries, too, might be able to use locally generated technologies that make the best use of domestic natural resources and abundant labor.

These elements produce a profile of an economic distribution and production structure different from those of the developed countries. One consequence of this is that countries in the region must behave differently to recover from recessions and other economic crises.

At present, a key component of economic recovery in the developed countries is the replacement of marginal production equipment with more innovative equipment. Companies do this to stay competitive in markets; to open up new domestic and foreign markets; and to stimulate the consumer to buy through lower prices, higher quality, or modified goods. For the countries of the Latin American and Caribbean region, this solution and the resumption of economic growth are hampered by two main difficulties:

1. Investment is curbed by an economically depressed environment with little or no expansion of foreign and domestic demand, uncertain economic and financial policies, and inflation. Some of these problems would tend to improve and others to worsen with the adjustment measures implemented or proposed for the debt crisis. These measures include the "Baker plan" and International Monetary Fund economic adjustment techniques suggested to debtor countries as conditions for backing new loans.

2. The tight domestic market removes stimuli from productive investment, and this is combined with weakness in the demand for traditional export products. Moreover, the limited domestic capacity for producing capital goods hampers solving these problems through massive and cheap production of products for the domestic market or the penetration of larger foreign market areas by new nontraditional exports such as semimanufactured products and electronic components.

Thus, the measures that Latin American and Caribbean countries take to solve crises must consider both economic elements and their structural roots. Short-term measures must be compatible with the long-term strategy and must contain the following elements:

- policies aimed toward greater participation in the world economy as well as domestic development;
- transformation of the production sectors and their technical bases to enable them to generate self-sustained economic growth and support social development;
- investment and consumption to improve income distribution; and
- expansion of production and employment through technical change.

LINES OF ACTION

Given the technical requirements for achieving sustained economic growth, a better international trade position, and lower unemployment, there is a clear need to consider technology in any action plan. This raises a series of technology-related questions: How do we encourage the production sector to apply the output of research and development to innovations in production? What changes would be necessary to make a particular production sector compatible with the region's characteristics? What technological changes would be needed to preserve the region's current comparative advantages and to develop new ones in new product lines?

To deal with these questions, any action plan should address the following areas:

- development of local technological capacity;
- internal and external transfer of technology;
- strategic projects that integrate science and technology; and
- government policies on science and technology.

Development of Local Technological Capacity

Internal technological capacity has been defined as the capacity of a country to select, acquire, generate, and apply technologies to attain its development objectives. Technological capacity is not an absolute concept, but refers to specific national objectives. It is a means, not an end. Thus, the technological capacity needed to produce an abundance of food for the population while also promoting the highest possible level of employment in such production differs from what is needed to explore outer space. Technological capacity unmatched to economic and social objectives is of limited usefulness (Bhalla and Fleitman, 1986).

Indicators that can be studied to evaluate and enhance the technological potential of a region or industry include:

- the professional qualifications of employees;
- the technical level of production;
- product quality; and
- the availability of trained workers and equipped laboratories.

Any action plan should establish goals for local technological capacity as well as the integrated development of society as a whole.

Another way to study the problem is to think of it as a tridimensional matrix. The first dimension would include the various components of technological capacity—the capacity to select technology, negotiate its transfer, and adapt and generate technology. The second dimension would include the factors that shape technology: human, financial, or information resources; plans and policies; institutions and infrastructure; and the natural, cultural, and social environments. The third dimension would cover objectives: enhanced competitiveness of national output, job creation, increased food security, import substitution for specific goods, and the like (Bhalla and Fleitman, 1986). This methodology will reveal the technological capacity available to achieve the country's objectives and the weaknesses that need to be remedied.

Several international organizations carry out activities to support the development of Latin America's technological capacity. These include the Organization of American States (OAS), through its Regional Scientific and Technological Development Program and its multinational, national, and special projects; the Inter-American Development Bank (IDB), through various lines of credit; the United Nations Development Program (UNDP), through specific national projects; the Board of the Cartagena Agreement, through the Andean Technological Development Program; and the International Research for Development Center of Canada. The Regional Scientific and Technological Development Program of the OAS has oriented its projects toward increasing technological capacity. Despite limited resources, it has been active throughout the hemisphere, responding to the interests of smaller and relatively less developed countries, through science and technology projects linked to their specific degree and form of development—such as aquaculture and the use of land and biotic local resources with economic potential.

In the future, any action by these organizations should be considered in view of the following issues:

- Changes in technological development since the program's inception may require that additional resources be allocated.
- A dual strategy is needed: To improve cooperative mechanisms to meet the new technological needs of the region while also maintaining efforts that have succeeded over the past 15 years.
- All of the actions proposed under the plan should prescribe that technological management be local. This requires a thorough inventory of the resources and needs of each business and organization.

Internal and External Transfer of Technology

Several countries in the region have already taken advantage of their capacity to produce new technology. They have succeeded more frequently when applying R&D results from institutes that specialize in the technological problems of particular sectors or serve large state enterprises, such as those for petroleum, electricity, and copper. Success was also achieved using R&D from parts of the scientific and technological infrastructure involved in carrying out large national programs.

Conditions for the successful transfer of R&D to the production sector include better-trained factory workers, wide application of standards and quality control systems, and access to current technical information. Positive experiences should be disseminated and broadened within countries and the region. This is a task for cooperation among countries.

Several constraints stand in the way of meeting these conditions:

- In the scientific infrastructure: Research areas that are not oriented toward solving the particular problems of a country, the unwillingness of scientists to address the urgent problems of the production sector, and the absence of institutional links between universities and the production sectors.
- In research and development institutes: Insufficient contact with the production sectors; flaws in the selection and management of projects; poor dissemination of research results; and in some cases, defective management. A priority in this field should be the training of administrative personnel in correct management techniques.
- In the sector of technological intermediaries: Shortages of companies and institutions to market and finance technology and insufficient development of national engineering and consulting firms, which are frequently biased toward the use of imported technology.
- In the production sectors: Lack of awareness of national technological capacity; strong aversion to risk, especially in technological innovation; and indifference to improving technical conditions of production.
- In the public sector: Economic policy measures that do not encourage the use of local technological capacity; flawed policy implementation

of scientific and technological development plans; insufficient coordination of research and development centers; and the need to improve resource allocation.

The transfer of external technology, on the other hand, faces different problems. The limited and sometimes highly bureaucratic approach that prevailed in the 1950s and 1960s must be changed to match the new global context. Revision is needed for two reasons. The first is the accelerating pace of international technical change and its impact on trade, which is greater since the recent recession in the developed countries. The second factor is that fundamental progress in the new technologies requires resources beyond what is generally devoted to science and technology.

Mechanisms to facilitate the transfer of external technology include:

- The establishment of channels and exchange markets to increase access to freely available technological knowledge, including that of developed countries.
- The promotion of regional consortia, with public and private participation that includes the scientific, technological, financial, and production communities; importers; and exporters. The consortia would encourage the manufacture of advanced technology products, drawing on the scientific and technological capacity and production of Latin America and the Caribbean.
- Modification of the most-favored-nation clause to include the particular characteristics of commercial exchanges of technology, and adapting norms of conduct agreed on internationally for foreign investment and the operation of multinational corporations to the conditions of the mixed consortia indicated above.

In summary, the plan would ensure that the transfer of technology between countries does not undercut local efforts to develop a technology infrastructure. Transfers within the region will create opportunities to substitute imports, to pool resources, and to expand markets. Correct management of the transfer of technology from outside the region could aid technical change, as long as unjustified social and economic costs are avoided.

Strategic Projects to Integrate Science and Technology

Projects in this area must pursue two strategies. At the regional level and at the national level of some of the larger countries, work should begin on a few key advanced technologies, such as biotechnology, microelectronics, and new materials. In countries with limited markets and less advanced science and technology infrastructures, activities should focus on adapting and developing only applications of advanced technologies, not on basic investigations of the technologies themselves.

This dual strategy requires study to see whether the region can compete in science with the rest of the world and whether a critical mass of trained people is available to sustain the development of science-based technologies, as well as technologies to exploit, conserve, and manage local resources of raw materials and energy. The countries in the region have been hampered in developing this cadre by the lack of financial resources to retain a significant number of high-level scientists, insufficiently developed postgraduate programs, and limited supplies of up-to-date laboratory equipment.

Nevertheless, in designing a program to develop a local capacity in basic and applied science, these factors should be taken into account: First, the region as a whole already has scientific capabilities that provide a basis for initiating the proposed projects. Second, pooling the scientific capabilities of several countries requires the cooperation of scientific centers with sufficient resources to ensure the international technical contacts of its personnel. Third, efforts should be made to carry out all aspects of technological development: research and development, experimenting in pilot plants, preparing for commercial application of technological innovation, negotiating of financial backing, and marketing the final process or product.

Government Policy on Science and Technology

In Latin America and the Caribbean, government is an important promoter of scientific and technological development. In particular, it plays an important role by setting in motion the mechanisms for financing scientific and technological development. The recent economic crisis and the need to advance technological change pose a challenge to the government's function in this field. As a result, government action on science and technology should consider and address two key issues. First, governments must emphasize the importance of democratic discussion of strategic options in science and technology. An example of this process is the public discussion in the United States on international competitiveness and technological strategies. Second, it is necessary for the government to refine its coordination of technological development programs and to support the necessary scientific and information base.

CONCLUSION

For the past 30 years, the development of the Latin American and Caribbean region has been characterized by external dependence and vulnerability on the one hand, and defects in the socioeconomic systems on the other. In the postwar period, various models emphasizing economic variables were promoted. These models have proved to be of limited use in dealing with the changing situation in the Latin American and Caribbean countries, and their use has not ended the region's socioeconomic heterogeneity nor reduced the dependence of the region's growth on the developed countries. Because sustained economic development in the region can occur only in association with positive social change, development policies must cover all of the main factors that make progress possible, including the effective use of technology.

REFERENCES

Bhalla, A. S., and A. G. Fleitman. 1986. Indicators and social and economic development. Science and Technology Series. Proceedings of the Panel of Specialists. Graz, Austria: Westview Press.

Brown, L. R., et al. 1986. State of World, 1986: A Worldwatch Institute Report on Progress Toward a Sustainable Society. New York: Norton.

González, N. 1986. Reactivación y desarrollo: el gran compromiso de América Latina y el Caribe. Presentation by the Executive Secretariat to the Twenty-First Session of the Economic Commission of Latin America, México, D.F., April 17–25, 1986.

Marí, M. 1986. Technology Gap: Support for Marginal Sectors and National Technology Policy. Washington, D.C.: Department of Scientific and Technological Affairs, Organization of American States.

Martín del Campo, E. 1983. Perspectivas y limites de la cooperación técnica y científica internacional en la década de los 80 y el caso de México. Washington, D.C.: Organization of American States. Mimeograph.

Norman, C. 1981. The God That Limps. New York: World Watch Institute.

Sagasti, F. R., and C. Cook. 1985. Tiempos difíciles: Ciencia y tecnología en América Latina durante el decenio de 1980. Lima, Peru. Mimeograph.

Schwartz, H. H. 1983. Bottlenecks to Latin American Industrial Development. Inter-American Development Bank study. Washington, D.C.

Waissbluth, M. 1983. Sinópsis de los problemas de la innovación tecnológica en América Latina. Latin American Meeting on Technology Innovation Management. OAS, SUBIN, FINEP, PACTO, São Paulo, Brazil, September 26–October 7, 1983.

World Bank. 1986. World Development Report, 1986. Washington, D.C.

Strategies for U.S. Economic Growth

RALPH LANDAU AND NATHAN ROSENBERG

WITH THE MARKED SLOWING DOWN of U.S. economic growth and the apparent decrease in ability to compete in an increasingly international marketplace, more urgent attention from many quarters is now being directed toward finding the causes and cures. In this paper, we review what is known about the impact of technological change on economic growth. We emphasize that technological innovation has been the historic engine of U.S. economic growth, and we argue that it is equally critical to future growth and competitiveness. We find also that high capital investment rates are directly related to high productivity growth, which in turn is the key to rising prosperity and preservation of a high-wage economy. We conclude further that technological innovation and capital investment are essentially two sides of the same coin, and that the one without the other will not contribute significantly to the nation's productivity growth. We analyze whether present U.S. economic policies, public and private, encourage or retard innovation and capital formation, and find that, in fact, they do not provide the stable, constructive environment required in these essential areas. Therefore, we propose improved strategies to preserve U.S. competitiveness and secure the social benefits of continued economic growth. We stress that competitiveness is not an end in itself but a means to increase such growth.

The growth rate in real income per person in the United States has been almost 2 percent per year since the Civil War. With this growth rate, standards of living nearly doubled between generations. Despite a simultaneous huge increase in population in this period, the United States moved from a largely rural economy to the greatest industrial power. Thus, the country's average real growth rate in gross national product (GNP) was about 3 percent per year; from 1948 until recently it attained 3.5 percent. The United States surpassed the United Kingdom, at one time the leading industrial power, which grew at a per capita increase of only 1 percent per year, and is now one of the poorer members of the European Common Market. On the other hand, Japan has surpassed even the high American growth rate in the period since the Meiji Restoration, which began in 1868. With a GNP growth rate of more than 5 percent since 1930, Japan has become the second-largest economy in the world.

Such is the power of compounding over long periods of time. Differences of a few tenths of a percentage point, which may not appear very significant in the short term, are an enormous economic and social achievement when viewed in the long term. Thus, it is of concern that the U.S. real GNP growth rate recently dropped to about 2.5 percent at the height of a long 5-year economic recovery. The basic question facing the United States today is whether it will be like the United Kingdom, while Japan and the Far East eventually outdistance it, or whether it will maintain a more prominent position of economic, and hence strategic, leadership.

QUANTIFYING THE ROLE OF TECHNOLOGICAL CHANGE IN ECONOMIC GROWTH

It is obvious that the United States could have achieved its growth in per capita income in either of two very different ways: (1) by using more resources or (2) by getting more output from each unit of resources. How much of the long-term rise in per capita incomes is attributable to each?

For many decades economists approached this issue of rising per capita incomes as if it were primarily a matter of using more resources, especially capital equipment, all essentially unchanged. The first serious attempts at providing quantitative estimates came during the 1950s, and the answers came as a big surprise to the economics profession.

When, in the mid-1950s, Moses Abramovitz (1956) and Robert Solow (1956, 1957), among others, looked at the quadrupling of U.S. per capita incomes between 1869 and 1953 and asked how much of the observed growth could be attributed to the use of more inputs, the answer was about 15 percent. The residual—the portion of the growth in output per capita which could not be explained by the use of more inputs—was no less than 85 percent. What seemed to emerge forcefully from these exercises was that long-term economic growth had been overwhelmingly a matter of using resources more efficiently rather than simply using more and more resources.

Abramovitz was himself very circumspect in interpreting his findings, calling it "a measure of our ignorance." Others attached the label "technological change" to that entire residual portion of the growth in output which cannot be attributed to the measured growth in inputs, and thus equated it to the growth in productivity. Productivity in this sense (multifactor) measures the efficiency of the inputs of both capital and labor, although the figures on productivity most frequently cited refer to labor productivity, and economists generally employ the term in this way. Strict economic interpretation of this residual is not entirely satisfactory, however. Many social, educational, and organizational factors and considerations of scale and resource allocation are also at work. Nonetheless, if the definition of the role of technological change is viewed broadly enough, it is probable that it is indeed the central component of productivity growth.

This awakened interest in the 1950s by academic economists in the causes of long-term growth led to other studies by economists like Edward Denison (1985) of the Brookings Institution and John Kendrick (1984) of George Washington University in the 1960s and 1970s who applied different techniques and obtained varying results from refining the component causes of such growth.

THE REALIZATION OF THE MAJOR ROLE OF CAPITAL INVESTMENT

Long-term real productivity growth rates conceal considerable erratic variation in the short run. This extreme variability is caused by extraneous shocks, such as the oil price rises, cyclical variations of the economy, wars, inflation, and many others. Nevertheless, since 1966 productivity increases have greatly diminished from previous levels. For the period 1960–1979, the multifactor productivity growth of the U.S. economy was only 0.26 percent per year, and in much of the later part of this period, the growth of total GNP was brought about almost entirely by increases in capital and labor, especially (in the 1970s) the latter as the number of baby boomers entering the work force peaked. Although the explanations for the collapse in American productivity have varied, it seems clear from extensive recent studies by Harvard's Dale Jorgenson and coworkers (1986, 1987) that the comparative performance of the U.S. and Japanese productivity growth rates (which in the same 1960–1979 period was 1.12 percent per year) has been heavily influenced by the much higher rate of Japanese capital investment. This high rate was well suited to a rapid adoption of the latest available technologies bought from many companies around the world, often at bargain prices. The Japanese investment was twice as high as the rate in many U.S. industries, which in some cases (e.g., steel) were not adopting technology with the same urgency. Other international evidence also suggests a high correlation between national investment and economic growth rates; for example, the Federal Republic of Germany and France, with investment rates roughly twice those of the United States, had about twice the productivity growth rate. Jorgenson's research suggests that in the postwar era, capital formation has accounted for about 40 percent of economic growth, with productivity growth being 30 percent and increase in labor also being about 30 percent.

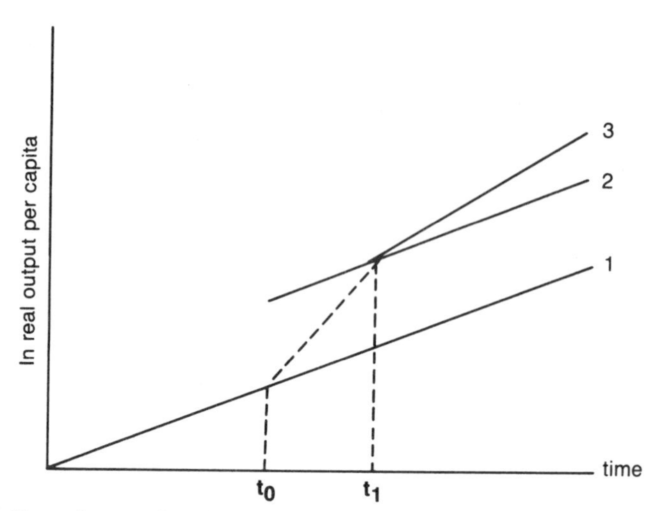

FIGURE 1 Alternative growth paths—technical change and capital formation. t_0, Proinvestment policy leads to higher capital formation and transition to higher level of income. t_1, Economy resumes long-term growth rate, or through interaction of investment and technical change, moves to more rapid growth path. SOURCE: Boskin (1986, p. 37).

Assar Lindbeck (1983) at the World Bank and Jorgenson et al. (1987) have each made recent extensive studies of the possible reasons for the productivity slowdown among various nations, and both draw particular attention not only to the lower rate of capital accumulation but also to the need to reallocate resources, because of the rise in energy costs or because of new regulations for greater environmental and health protection, etc. That the latter has had a detrimental effect on American international competitiveness, even though conferring genuine domestic benefits, is shown in a careful economic study by Joseph Kalt (1985) of Harvard University.

HOW GROWTH RATES CAN BE INCREASED

A complication in understanding the causes of growth is that quantitative measures of productivity do not fully describe the performance of any economy. Quality is difficult to measure, and of course, many goods and services have greatly improved in performance and variety and many new products have appeared, whereas the quality of other services may have deteriorated.

Nevertheless, significant insight into the key relationships is suggested by Figure 1. According to Michael Boskin (1986) of Stanford University, the fundamental variables that increase the rate of growth of a country in the long term are the rate of technical change (involving not only research and development [R&D] but also design, invention, market development, and the like) and the improvement in quality of the labor force. Merely increasing the rate of capital formation will lead to only a temporary increase in the rate of growth (moving from growth path 1 to growth path 2). This increase in capital formation occurred during the 1960–1979 period in Japan, although it has more recently declined. These difficult-to-sustain large growth rates, which occurred as international resistance to the export-led drive increased, have now diminished, but permanent advantages for many industries have been created.

Measurements of productivity growth alone are not, however, a complete free-standing expression of the role of technology in economic growth. R&D by itself is seldom performed unless it is expected to be employed in new or improved facilities and in superior operating modes. Technological change is thus heavily carried by capital investment, but it is also a powerful inducement to it, since the availability of superior technology is a major incentive to invest. Likewise, improvements in labor quality (human knowledge, skill, and training) are both a carrier of and a spur to technological change. Hence, within each of the basic factors of production, technology often takes an embodied form. When more business capital is invested, new technology is usually embodied in it—it is seldom just a pure copy of what went before. Thus, the capital stock of a country is of different vintages, reflecting such changes.

Boskin's growth path 3 in Figure 1 illustrates the point that if these interactions are in fact occurring, then the rate of investment can move the economy to a longer-term higher growth rate, as opposed to a mere upward shift in the level at any given time. This is especially true for the results of "breakthrough" R&D, which require large new investments. A substantial proportion of R&D is devoted to improving the performance of capital in place, or for ready modification or retrofitting by incremental investments, which require less new capital. Often even maintenance capital may embody improved technology, although the constraints imposed by the total process and product eventually force decisions for complete new plants using later technology if the firm is to remain competitive. Hence, in view of the really novel technologies now becoming available and the effects of continuing R&D efforts, the need for totally new facilities and closing down of obsolete units is becoming much greater if the United States is to remain competitive and to catch up with the Japanese spurt of the last 20 years. Some plants and even companies will disappear in the process. The interaction of investment and technological change constitutes a major guide to future policy. The earlier distinctions between capital investment and technological change as separate causes of growth need to be modified in favor of a view that sees them as largely parts of the same process. It is in this broad sense that technological change probably has been the main cause of as much as 70–80 percent of U.S. economic growth.

Technological innovation involves managing the reduction of uncertainty—both technical and commercial (economic). It is for that reason that engineers, the agents of technological change, require, in addition to a sound science base, abilities in design, management, and economic judgment. It seems evident from the worldwide slowdown since the early 1970s that managements and engineers everywhere have been confronted by much greater uncertainty about the business environment and have been compelled to expend much effort in dealing with it. This uncertainty included not only the energy crisis but also adverse macroeconomic policies by many nations, the collapse of the Bretton Woods fixed currency exchange rate system, the Vietnam War and its consequences, inflation and inflationary expectations, new sources of competition, and other changed circumstances from the relative tranquility of the earlier postwar years.

Thus, to improve the erratic and unsatisfactory growth rate of the U.S. economy, the major prerequisites are increased capital investment of constantly improving quality, increased training and skill of the total work force, and increased investment in R&D. These all interact with each other and are complementary. But they will not be easy to obtain without a favorable economic climate for long-term steady growth.

THE WORLD HAS IRREVOCABLY CHANGED

A most notable change in the past two decades has been the extraordinarily rapid diffusion of technology to many other countries, including those with wage rates a great deal lower than those of the United States. A century ago, transfer of technology from one country to another took place over many years; today it may be a few years or even months. But now that industrial technology has taken root in many new places, many countries no longer depend on the technical progress of just a few industrialized nations; they have become serious new competitors. This is a historic upheaval.

Such diffusion and its consequences pose immense new difficulties as well as opportunities for the United States. It has posed particular problems for managements brought up in the earlier postwar years when the United States had the world markets as well as a pent-up domestic market to ensure ample aggregate demand for anything that could be made—even cars with no more than novel tailfins. But today's managements, particularly in the manufacturing sector, face quite a changed world scene. Many, but by no means all, have coped quite well, such as by spreading their operations across international boundaries, and thus insulating themselves against the varying national policies that affect operations in particular countries. The recent cost-cutting efforts of many managements display a sharp awareness of the challenges from competitors. Such efforts suggest the presence of significant long-run strategic vision of how their companies must grow and compete. They must now do even better. Those companies that have not or do not recognize this will either be taken over or disappear. Indeed, in the past, the successful entrepreneurial exploitation of new technologies to create new products, processes, and businesses has been *America's distinct comparative advantage*. Now, the greater number of international competitors poses greater challenges to the United States.

Markets for goods and services are now global. Many countries are competing vigorously with the United States in advanced technologies such as electronics, space, nuclear power, chemicals, and autos, producing products that often are of better quality, greater reliability, and lower price. In addition, financial markets penetrate everywhere and impose their own discipline on countries' policies. Trading in financial assets far exceeds trade flows in goods and services (perhaps more than $50 trillion per year versus $3 trillion). The financial system of the world has become international, while the individual monetary, industrial, fiscal, trade, and labor systems remain national.

American wage rates have traditionally been high, although exact comparisons depend significantly on the exchange rates, productivity rates, and inflation rates. Many countries such as Brazil, South Korea, Mexico, and Taiwan still have much lower labor costs and are increasing their productivity faster in a catch-up process. If the United States is to maintain a higher-wage economy, it must raise its growth rate, its productivity, and its rate of innovation. Clearly, the United States is no longer a largely isolated economy and has lost the power to control its destiny virtually unilaterally. Its growth rate now depends to a much greater degree on its ability to compete successfully in international markets, especially in manufacturing, the most exposed.

THE KEY ROLE OF MANUFACTURING

Despite the fact that manufacturing now employs only 20 percent of the work force and has maintained over the long term about a 22 percent share of real GNP, it is exceptionally important to the U.S. economy because:

1. It performed about $85 billion out of $88 billion spent by the total private business sector in R&D in 1986, virtually all of the applied R&D of the United States.
2. It provides the major component (two-thirds) of the foreign trade of the United States, which consists of about 8 percent of GNP as exports and 12 percent as imports. The combined value of these two figures was only 8 percent in 1970. The same handful of companies, primarily those exposed to foreign competition, dominate both R&D spending and exports (Table 1), and they are, on the whole, among the major investors for increased competitiveness.
3. The manufacturing sector purchases a large part of the output of the service sector, which reciprocates.

TABLE 1 America's Leading Exporters and R&D Spenders, 1986 (in order of size)

Exporters[a] (1986—Over $1 Billion)	R&D Spenders[b] (1986—Over $500 Million)
General Motors	General Motors
Boeing	IBM
Ford	Ford
General Electric	AT&T
IBM	General Electric
Dupont	Dupont
Chrysler	Eastman Kodak
McDonnell Douglas	United Technologies
United Technologies	Hewlett-Packard
Eastman Kodak	Digital Equipment
Caterpillar	Boeing
Hewlett-Packard	Chrysler
Allied-Signal	Xerox
Digital Equipment	Exxon
Philip Morris	Dow
Occidental Petroleum	3M
Union Carbide	Monsanto
Westinghouse	Johnson & Johnson
Motorola	McDonnell Douglas
Raytheon	
Archer Daniels Midland	
General Dynamics	
Total: $56.8 billion U.S. merchandise exports	Total: $23.59 billion R&D expenditures

[a]SOURCE: *Fortune*, 20 July 1987, pp. 72–73.
[b]SOURCE: *Business Week*, 22 June 1987, pp. 139–159.

4. Productivity improvement in manufacturing has been historically higher than that of the service sector, contributing heavily to the overall growth and wealth of the economy and helping hold inflation down.
5. This sector produces material essential for national security.

It is because of these features that concerns exist about the ability of U.S. manufacturing firms to remain competitive, especially in the case of the many small firms that contribute so much to manufacturing. Nevertheless, spurred by this competition, American manufacturing has been improving its productivity markedly. Labor productivity grew by more than 3 percent per year between 1980 and 1984, and is still growing at close to that rate, in contrast with a 1.5 percent rate in the 1973–1980 period. True, Japan experienced a 5.9 percent productivity growth rate during the 1980–1984 period, slipping to a 5 percent rate more recently; but many American firms are responding to the competitive conditions, aided by U.S. economic policies of the recent past, which also partially favored new investment.

THE SIGNIFICANCE OF THE NEW TECHNOLOGIES TO FIRMS

A highly significant long-term trend is the increasing impact of the information age on the entire economy, which is changing the nature of manufacturing itself. In fact, this new information-processing industry, when joined with the new technologies of automation, represents at least as fundamental a change in the economy as electrification did in the early twentieth century. It may well be the precursor of a new Industrial Revolution. The impact of this new industry on the rest of the economy is only beginning to be felt, and it will doubtless be some time before its effect on productivity improvement is fully realized. Such lags are not at all unusual in the case of such fundamental innovations. Thus, in the early decades of the twentieth century, business sector productivity failed to grow significantly, at a time when many technological changes were occurring. Not until the early 1920s did the productivity increases begin to occur. The introduction of the assembly line and the electrical motor produced 20 years of upheaval in industrial organization and management strategies. It was the combination of associated institutional changes (including educational, legal, and financial systems) and entrepreneurial creativity that eventually led to rapid productivity growth in the business sector. In the process, not only institutions and organizations had to change; a great deal of existing capital was rendered obsolete. The energy cost changes in the 1970s had a similar effect.

Coupled with this transition are the new technologies now becoming available. The next great wave of technological innovation is rolling in—in physics-, chemistry-, and biology-based industries. These new technologies can be rapidly employed to introduce higher value-added products and services, which is a sensible prescription for America in a world that offers increasing competition from economies with much lower wage rates. Higher value added means higher value added per worker. The point is that sectors where each worker has a high value added are sectors where there are extensive inputs of capital per worker—both tangible and nontangible capital. In such sectors even a high wage still constitutes only a small part of the total cost of production. Only in this manner can the outflow of U.S. jobs be arrested and reversed while the standard of living is improved.

There is one essential prerequisite for an economy to succeed with a strategy such as that proposed here: Capital in all its forms must be abundant, cheap, and of increasing quality. This in turn means pursuit of policies that encourage high rates of savings and capital formation and enhanced education, training, and research and development. This, as noted above, is exactly what the Japanese have been doing, and they have been doing it much better than the United States has. These prerequisites can be recognized as the key components of Boskin's (1986) prescription for sustained growth.

In summary, the American manufacturing industry must now move even more rapidly into higher value-added, more technologically sophisticated products made by more technologically sophisticated processes within more flexible organizational structures, and away from commodities that can be made more cheaply abroad, if American living standards are to be preserved and improved. American manufacturing must become more competitive by becoming more technological, which also means more capital- and skill-intensive; and the service sectors, especially those connected to information technology, must do the same. Commodity manufacture can be successful, however, where the large domestic market permits the use of favorable plant scale, especially if coupled with advanced technology. On the other hand, flexible automated manufacturing may reduce the economies of scale and aid smaller companies in their ability to compete. Even where the bulk of the value added is in marketing, sales, and distribution, large amounts of capital are still required with close feedback to manufacturing. R&D-intensive companies are themselves capital intensive, because R&D is a long-term capital investment.

This formulation of the American imperative reflects a basic shift in international trade from the situation where countries and industries could count on long-term stable comparative advantages to a dynamic state, in which comparative advantages are constantly being altered.

THE ROLE OF GOVERNMENT

There is still vast ignorance in American society about the forces which fuel growth, as well as antitechnology influences (especially noteworthy today in biotechnology). The public must be made aware of what really is at stake in a technologically intensive competitive world economy and, particularly, that competitiveness is a necessary but not a sufficient condition for economic growth. It is not an end in itself, but a means to improve long-term growth rates, the American standard of living, and quality of life.

What else can government do effectively? Government cannot decree a successful growth strategy, but it can better coordinate its various policies if it understands the real goals involved and promotes the infrastructure that the private microeconomy requires for innovation. The basic R&D function and the provision of safety nets for the unfortunate and those unwillingly excluded from the growth process are among these activities. Some regulatory activities are essential, although they can also be inhibiting. Foremost among the services that governments can provide is a better educational system. Although many skilled workers of today may not be able to adapt well to the new technologies, the younger generation can be trained to be far more adept at employing them. But continuing education is becoming more important everywhere.

THE MACROECONOMIC ENVIRONMENT FOR GROWTH AND COMPETITIVENESS

Although there is good reason to be skeptical about the ability of government to make intelligent decisions in specific markets, there is a great deal that government can do in providing a more favorable environment for business decision making. If the roller coaster economic experience of the last 20 years has proved anything, it is that long-term real growth is brought about by change and improvement in the microeconomy. A macroeconomic program, directed at controlling economic fluctuations by fiscal and monetary policies, primarily affects short-term growth. Nevertheless, some macroeconomic policies, such as high interest rates and therefore high cost of capital, may have a long-lasting impact on the microeconomy, for better or worse. It is important to recognize this interaction in setting macroeconomic policy, so as to improve long-term growth prospects. This is particularly important with respect to the volatility of macroeconomic policies.

The complementary and interrelated nature of capital investment used by well-trained people and based on the latest technology has been argued in this paper. Unfortunately, the American savings rate from which such investments can be made is low in comparison with the rates of its principal trade competitors in East Asia. Boskin has shown that the savings rate of the Japanese, depending on how it is measured, is from two to nearly three times that of Americans (Boskin and Roberts, 1986). It is estimated that net domestic U.S. investment in 1986 was funded only 50 percent by net national savings; 50 percent came from foreign capital inflows. Japan's high savings rate provides that country with a much lower cost of funds, permits heavy domestic investment as well as export of capital to other countries, and a "patient money" approach toward investment and R&D. If the cost of capital in Japan is half of that in the United States, as it appears to be—and it is much more abundant—then their horizon for decision making can be twice as long! This is the real *Japanese comparative advantage,* and is exemplified in Figure 2, which shows the resulting high productivity growth. This high savings rate is a postwar phenomenon and was carefully designed by encouraging tax-free savings and other measures. In addition, Japan has a high educational base and a number of unique societal institutions that shelter firms against risks and allow them to undertake large, long-term projects.

By contrast, postwar U.S. economic policies have generally favored current consumption over investment, which is likely to result in the reduction in the rate of increase of consumption and standard of living of future generations. In recent years, this has been accomplished by a fiscal policy of

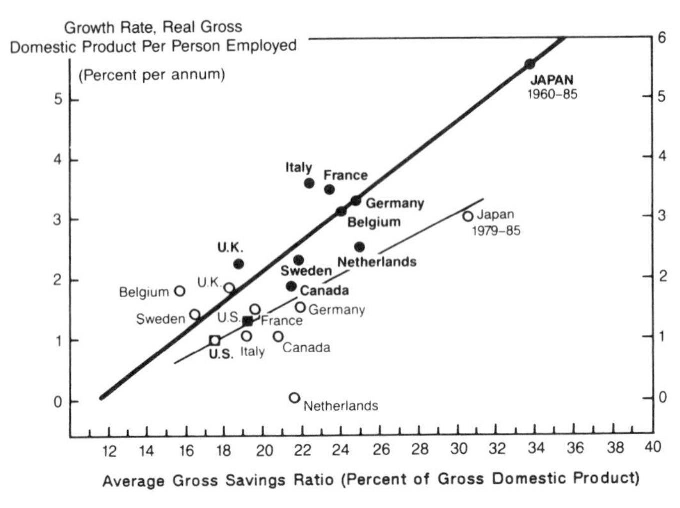

FIGURE 2 National rates of savings and productivity growth, 1960–1985. Gross domestic product includes personal, business, and government savings. Symbols: •, 1960–1985; ○, 1979–1985. SOURCE: Organization for Economic Cooperation and Development (Courtesy of U.S. Department of the Treasury) (1986).

large government deficits (as much as 5 percent of GNP), accompanied by a tight monetary policy that has produced high real interest rates and a high cost of funds. The dollar consequently became overvalued and converted a small overall trade balance in 1981 to an annual deficit of perhaps $140 billion currently, injuring numerous American companies and industries. Although the dollar is much weaker against the Japanese yen and the German mark since the intervention by the finance ministers of the five major Western countries, it is still almost unchanged against the currencies of many American trading partners, such as those of Canada and South Korea.

Debt has accumulated at all levels at an accelerating pace, so that it now costs companies nearly 40 percent of cash flow to service debt versus 30 percent a decade ago and less than 20 percent in the 1960s. Interest on government debt has risen from 7 percent of public spending in 1975 to 14 percent today. Consumer installment debt service is now nearly 18 percent of personal income versus 12 percent in 1975. Total public and private debt in 1986 was more than 200 percent of GNP, up from 163 percent in 1975 and 174 percent in 1980. Some of this debt has gone into mergers, acquisitions, takeovers, restructures, and leveraged buyouts. Most have the effect of substituting debt for equity, a tactic favored by the tax laws, which allows deductibility of debt interest by corporations but not dividends. The U.S. Treasury in effect pays the premium that the stockholders of undervalued companies receive. Sometimes the resulting company becomes more efficient; and restructuring, with or without takeovers, has produced major gains in efficiency and cost control in American manufacturing, contributing significantly to inflation control. More often, only the existing shareholders benefit (in the short run) and almost everyone else loses. The large debt cited above has not gone sufficiently into productive investment but has generated excess liquidity in the financial markets. Inflation in the 1980s is a financial phenomenon.

The United States is now an international debtor, and the longer the deficit continues, the greater the trade surplus will have to be in the future to finance the debt. With thinner cushions against economic downturn resulting from the increasing debt leverage, risk taking is discouraged or becomes foolhardy—not a climate conducive to gaining long-term competitive advantage or conducting long-range R&D or investment. The danger of recession and the consequent increase in the budget deficit is all the greater because government has already used up most of its fiscal and monetary stimuli. Furthermore, foreigners will at some point cease financing the American deficits. In 1986 the excess of spending over domestic production was more than 3 percent of the gross domestic product, and virtually all was financed by borrowing from abroad. This is unsustainable, and when foreign lending diminishes, the standard of living will inevitably decline: Capital for investment will become scarce because of the low net savings rate. Faced with such seemingly intolerable conditions, the government would be sorely tempted to monetize the debt, and with it, reduce its foreign debt while restoring price inflation as a potentially less painful way to reduce consumption. History suggests that, although there may be some favorable short-term effects, for example in employment, the long-term effects may be very unfavorable.

Science and technology, which require long time horizons, are bound to be undervalued in a short-term financial climate, which also favors financially oriented managers over engineers in promotions and salaries and short-term R&D projects over longer-term work, which might produce real breakthroughs. Inflationary expectations produce very short-term horizons indeed.

There was another side to this pro-consumption policy in the late 1960s and 1970s, when the surge of the baby boomers and women entering the job market crested. Because capital was relatively scarce and expensive and these new less-skilled workers were relatively cheap, the baby boomers and women flooded into the service industries, particularly into smaller companies, which play a vital role in new job creation. The result was an extraordinary boom in jobs (Table 2), wherein the United States did much better than its principal rivals. Europe still has much higher unemployment because of its greater labor and institutional rigidities. But, of course, the productivity of the American economy as a whole suffered by comparison, because policies were not in place to promote more of both jobs and productivity. In effect, American manufacturing remained competitive throughout the 1970s by keeping real

TABLE 2 Civilian Employment, in millions

Year	EEC[a]	EEC[b]		United States		Japan	
1955		101.4 (est.)		62.2		41.9	
			+3.4%		+14.3%		+12.9%
1965		104.8 (est.)		71.1		47.3	
			+0.5%		+20.7%		+10.4%
1975	121.8	105.4		85.8		52.2	
			+4.1%		+24.9%		+11.3%
1985	121.0	106.5		107.2		58.1	
1986	121.5[c]	107.1[c]		109.6		58.5	
Net increase:		5.7		47.4		16.6	

[a]12 members.
[b]10 members.
[c]Estimate based on 20–25 percent of total.
SOURCE: Organization for Economic Cooperation and Development and European Economic Community.

wages down and reducing profit margins, thus compensating for the lack of new, and the lower cost of, production capital investment. The corresponding inflation masked the declining competitiveness because prices were raised by firms faster than real wages were adjusted. Even so, the last year of a positive merchandise trade balance was 1975; it averaged a negative $40 billion or so until 1982, when it began its plunge to a record $170-billion deficit in 1986.

So, from the early 1980s on, the American manufacturing strategy of the 1970s was ended by the strong dollar; firms were exposed to world deflationary trends and could not mask inefficiencies. The process of recovery, which entails adopting new technologies to regain a competitive edge, will be both expensive and perhaps even destructive of some jobs, of companies, and of certain industries that cannot realistically expect to restore their lost competitiveness.

However, the demographics are also changing—the baby boom is over. The labor force grew at a rate of 2.9 percent in 1976–1980, but currently it is growing at 1.8 percent, and the rate of growth is declining. Meanwhile, managements invested heavily for almost two decades in information technology, which will be required as skilled labor becomes scarcer and more expensive, rather than in assembly line capital. As a result, the ratio of capital to labor rose rapidly in the information sector, while capital per industrial worker stagnated. Nevertheless, the effect of information technology investment shows up in the recovery of manufacturing productivity in the 1980s. The service sector should follow.

The time has come, therefore, for macroeconomic policies that would favor large new investments in all sectors, both for replacement of obsolete facilities and for installation of the new technologies that would favor increased productivity growth while still providing for an adequate number of new jobs. Because foreign capital inflows of the past must be paid for by exports, this policy will require a major increase in U.S. savings. A trade gap means that a country spends more than it earns, which is identical to saying that it invests more than it saves. If prudent noninflationary long-range macroeconomic policy is to be followed, this would require a tighter fiscal policy (reduced dissaving by governments) and a looser monetary policy to supply the money for increased growth at lower interest rates, combined with sensible regulatory and legal policies and a tax system that favors investment. Such a tax system would more nearly resemble that of the Japanese in allowing essentially tax-free saving and lowering the cost of funds. Most economists concur that a consumption tax system of suitable progressivity would favor a higher savings and investment rate. The 1986 Tax Reform Bill is not such a system and will in fact probably increase the cost of funds. This is a disastrous trend when capital and its cost have become the principal factors in international comparative advantage and competitiveness. As noted above, this is the other side of the coin of technological innovation. The current talk of reducing the budget deficit by raising taxes begs this question; if new taxes merely reduce savings in the private sector while producing an equivalent gain in savings in the public sector, nothing fundamental has been accomplished.

Most probably, to avoid the risk of recession if the budget deficit is brought down too hastily, and of inadequate savings if the trade deficit is sharply reduced, both need to be addressed simultaneously. A beginning has been made, and there is a modest downward trend in the twin deficits as a percentage of GNP. If continued, a cheaper dollar could eventually reduce the trade balance deficit, while the net national savings rate would improve as government dissaving is reduced. The reduced trade deficit would permit domestic demand to replace the loss of demand brought about by lower government deficits, as would increased demand for American exports abroad. Finally, steps can be taken to improve the savings rate in the private sector, which is a much longer-term but essential problem. Such a "soft" landing would aid the entire world economy to continue to grow; in this way, the nations are not deadly rivals by virtue of productivity increases and they permit the world economy as a whole to grow—a *positive-sum strategy*.

If the United States falters in following such a benign pro-technical-change policy, it may fall into varying degrees of protectionism, in addition to the perils of inflation. The world's military and financial stability may be undermined. The burden of low productivity growth is forced upon the consumers. The consequences will inevitably lower the standard of living and security of the American people, as it is a *zero, or negative, sum game*.

Other countries are sure to retaliate. The United States may then retrace the path of Great Britain. Although there is certainly room for selective trade actions to meet egregious examples of foreign protectionism in order to improve overall world trade and therefore growth, "managed trade always ends up being managed on behalf of special interests," not the general public, as Paul Krugman (1986) of the Massachusetts Institute of Technology has stated. The purpose of such selective trade actions should be to open up other nations' markets, rather than close those of the United States. Recently, the United States has been acting more vigorously to overcome perceived ingrained protectionism abroad and to counteract "dumping" at home, and has come to the conclusion that it can no longer afford to be the principal world locomotive of growth. The dividing line between such actions and counterprotectionism is a very fine one, however, and there are dangers inherent in such steps if injured interests can politicize the process.

Many in government are already using the concept of "competitiveness" as a code word for old-fashioned government intervention in industry, under which bureaucrats, politicians, and pressure groups try to protect failing companies and promote supposed "winners," while in fact promoting mostly each other's interests. It has also been widely noted that American institutions make it difficult to have a full-fledged industrial policy as advocated by some, but that may be a blessing in disguise. Government regulatory agencies tend to become the captive of special interests through the political process. The rapidly changing conditions imposed by the global pace of technological innovation will make it even more difficult for government to control or direct the private sector, and the interventionist role of the Japanese government in this regard has been exaggerated. As Daniel Okimoto (1986) of Stanford University says: "What lies within the [Japanese] government's effective power is largely limited to the creation of a healthy environment for business growth. . . ." If this is true for a Japanese government operated largely by a powerful, prestigious bureaucracy, under an essentially one-party government where the Finance Ministry has always been much more powerful than the Ministry of International Trade and Industry, how much more true is it of the American system of constitutional law, with a strong and increasingly intrusive court system, in a country with a largely powerless bureaucracy and two contentious political parties, and where the Secretary of the Treasury has always outweighed the Secretaries of Commerce and Labor?

In many countries, managed trade is no problem—the business interests and government tend to see things the same way. In the United States, that is not the case. Companies would soon lose touch with global technological and economic developments; LTV's bankruptcy shows the damaging effects on the steel industry of the trigger price system. The absence of foreign competition would reduce the pressure on companies to excel and develop.

Protectionism might even extend to capital flows, as happened in the 1960s, which could seriously harm American growth and lead to a decline in the stock market, which has been buoyed by the inflow of foreign capital. Maintenance of American competitiveness in the long run depends on sustained increases in productivity. Effective actions to achieve this will be required in both the public and private sectors. For example, in addition to those measures already described, technologists in private companies may well be asking themselves whether new high value-added technologies can be developed that are not as capital-intensive as present technologies—perhaps even more knowledge based.

So, a *positive-sum strategy* boils down to this: Only if Americans become better trained and managed—and invest a great deal more capital and technology in both manufacturing and services—can the standard of living improve at an acceptable rate in a highly competitive world market. The United States has some real historically demonstrated advantages in such a competition, but it must take a longer-term view and pursue those seemingly unexciting few-tenths-of-a-percentage-point increases in growth rate each year.

REFERENCES

Abramovitz, M. 1956. Resource and output trends in the United States since 1870. American Economic Review 46:5–23.

Boskin, M. J. 1986. Macroeconomics, technology, and economic growth: An introduction to some important issues. P. 35 in The Positive Sum Strategy: Harnessing Technology for Economic Growth, R. Landau and N. Rosenberg, eds. Washington, D.C.: National Academy Press.

Boskin, M. J., and J. M. Roberts. 1986. A Closer Look at Savings Rates in the United States and Japan. Working Paper No. 9. Washington, D.C.: American Enterprise Institute.

Denison, E. F. 1985. Trends in American Economic Growth 1929–1982. Washington, D.C.: The Brookings Institution.

Jorgenson, D. J., M. Kuroda, and M. Nishinizu. 1986. Japan–U.S. Industry Level Comparisons 1960–1979. Discussion Paper 1254. Cambridge, Mass.: Harvard Institute of Economic Research.

Jorgenson, D. J., F. M. Gollop, and B. M. Fraumeni. 1987. Productivity and U.S. Economic Growth. Cambridge, Mass.: Harvard University Press.

Kalt, J. P. 1985. The Impact of Domestic Environmental Regulatory Policies on U.S. International Competitiveness. Discussion Paper E-35-02. Cambridge, Mass.: Harvard University Press.

Kendrick, J. W., ed. 1984. International Comparisons of Productivity and Causes of the Slowdown. Cambridge, Mass.: Ballinger Publishing Company.

Krugman, P. 1986. A trade pact for chips. New York Times, August 10:F-2.

Lindbeck, A. 1983. The recent slowdown of productivity growth. Economic Journal 93:13–34.

Okimoto, D. I. 1986. The Japanese challenge in high technology. Pp. 541–567 in The Positive Sum Strategy: Harnessing Technology for Economic Growth, R. Landau and N. Rosenberg, eds. Washington, D.C.: National Academy Press.

Organization for Economic Cooperation and Development. 1986. Historical Statistics. Paris.
Solow, R. 1956. A contribution to the theory of economic growth. Quarterly Journal of Economics 70:65–94.
Solow, R. 1957. Technical change and the aggregate production function. Review of Economics and Statistics 39:312–320.

Improving the Quality of Life Through Technology

HAJIME KARATSU

In a recent Harris poll, American manufacturing executives were asked which countries would pose the most serious competitive threat to American manufacturing over the next 5 years and in the year 2000 (*Business Week*, January 12, 1987). Sixty-nine percent of the executives answered that emerging countries such as Brazil, South Korea, and Taiwan would be America's chief competitors in the manufacturing sector over the next 5 years, and 76 percent responded that those countries would pose a threat to U.S. manufacturing in the year 2000. Only 29 percent of those who responded said that Japan would be a serious competitor over the next 5 years, and only 11 percent cited Japan for the year 2000.

These findings illustrate an interesting trend. In the past, a nation's competitive power was determined by its geographical size and population. Beginning in the eighteenth century, however, the industrial revolution changed the balance of power among nations, and today even a small nation can achieve affluence and economic strength through its industrial achievement.

In Southeast Asia, there are major differences between nations even though they are located in the same geographical region and are surrounded by similar natural conditions. Singapore has a high wage rate, second only to Japan in Asia, yet it is a tiny island comparable in size to Manhattan and has a population of 2.7 million people. It is also located in a tropical zone with few natural resources.

On the other hand, the people of other nations in Asia are still living at a primitive level. When we look carefully at statistical data on the status of each nation, we see the correlation between a country's economic standing and various indicators of the quality of life. High economic figures are indicative of the advanced state of industrialization of each nation.

COPING WITH CHANGE THROUGH TECHNOLOGY

Just before the oil crisis of 1974, *The Limits to Growth,* a controversial report prepared for the Club of Rome, projected a very pessimistic scenario for the future of the global economy and industrialized nations in particular (Meadows et al., 1972). Given the atmosphere at that time—which included a general economic slowdown and the antipollution and antitechnology movement—the report had an enormous impact. It supported the theory, and more importantly the prevailing mood, that the global economy was headed for a period of decline. These influences contributed to a certain pessimism in many industrialized countries.

However, nations have demonstrated that they can cope with such conditions through the creativity of human beings. Faced with the oil crisis, Japan introduced innovative energy-saving technology into the steel industry, and today not a drop of oil is used in that sector. Japan has achieved an increase in its gross national product (GNP) of 2.7 times that at the time of the first oil crisis, while oil consumption has decreased to 80 percent of that in 1974.

Almost every industrialized nation instituted similar energy-saving measures. These efforts to eliminate energy losses in factories, automobiles, and elsewhere were successful in overcoming the energy price hikes. As a result of the new technology, decreased oil consumption has even forced oil producers to cut the price of oil.

Pollution in the industrialized areas of Japan, a by-product of the push for high economic growth, was another major problem in the 1970s. However, after a radical antiindustrialization movement became active, the Japanese government issued numerous antipollution laws. The strictest automobile emission regulations in the world were instituted in Japan in 1975, and cars that did not meet the emissions control specifications could not be sold in Japan. Such regulations were applied not only to automobiles but also in every factory. Consequently, the engineers working in the regulated sectors made great efforts to develop technologies within the framework of the new constraints.

As a result of those efforts, the air and water of Japan today have become clean again. It is said that half the budget to construct new ironworks plants was spent on energy-saving and antipollution devices. The average expenditure of the energy-saving/antipollution industry, which did not exist in Japan before the 1970s, is estimated to be $15 billion per year. Recently, these energy-saving and antipollution technologies have begun to be used all over the world, especially in Western Europe to eliminate pollution caused by acid rain.

INCREASING THE ECONOMIC PIE THROUGH TECHNOLOGY

The importance of a strong manufacturing base and the economic advantages of industrialization are well illustrated by Japan. Japan is one of the world's most crowded countries. With 2.7 percent of the world's population yet only 0.3 percent of the land area, Japan has few natural resources and is located on the fringe of the Asian continent, which is far from the world's main markets. Yet in 1986, Japan achieved a GNP of $2.3 trillion, 11 percent of the world's economic activity. The locomotive force of the Japanese economy is clear. It is technology. Many attempts to understand the basis of Japan's success, however, are marked by misconceptions. Some commentators say Japan has merely followed in the path of Western Europe and the United States or imitated ideas from developed nations and in this way moved ahead in manufacturing and other technologies. Some of these statements may be true, but my experience in the Japanese manufacturing sector since World War II has provided an insight into two key factors of Japan's success. The first factor is the way Japanese manufacturers develop new products through innovative technology. The second factor is the way the Japanese cope with and overcome problems that occur on the manufacturing shop floor.

In October 1985, I attended a conference in Toulouse, France, on advanced technology. During the conference, I wondered whether many Western Europeans understood the real meaning of advanced technology. When new technologies appear in the world, Western Europeans tend to apply them in complicated ways such as in space technology or missiles. Since these are difficult fields they seldom apply the advanced technologies in immediately practical ways. On the other hand, the Japanese make use of new technologies in whatever form seems to be easily applicable at the time.

Consider carbon fiber, for example. It is a highly innovative new material, lighter than aluminum and stronger than steel. Japanese manufacturers first used it for the shaft of golf clubs. Next they used it for fishing rods. And because they were using these new materials for simple products, even if some minor defects occurred, serious problems were avoided. After they perfected these production techniques for carbon fiber, Japanese companies used carbon fiber in more complex applications.

A more recent example is that of shape-memory alloy. In Japan, manufacturers started using this alloy in every possible field and explored many different product areas—such as air conditioners, eyeglass frames, and coffee makers. Consequently, Japan produces more of this alloy than any other country, 90 percent of the world total.

The most important strategy for using innovative technology is discovering and developing a new, profitable market. Technology should not stay at the idea stage; it should be converted into marketable products. Japanese firms are successful at commercializing new technologies because they select technologies with ready applications and move quickly in developing and manufacturing the product. A driving force in maintaining this commercialization strategy is severe competition among Japanese firms.

Another point is the difference in assumptions between Western European and Japanese engineers. If I talk with Western European engineers, their discussions tend to be "digital." They always think in terms of black or white and yes or no. This Cartesian way of thinking was quite effective in the natural sciences, where greater simplification is necessary to organize ambiguous data. However, production activity is not that simple.

Basically, manufacturing is a battle against thousands of different possible breakdowns and errors: mistakes in planning schedules, incorrect design, accidental mixture of materials, and so forth. Moreover, machines do not always work uniformly, and factory workers occasionally make mistakes. If these errors accumulate, the result will be a pile of defective goods. The lesson here is that it is easy to fail if you are not aware of all the "gray areas" of production.

We cannot predict where and how such errors will occur, however. Everyone in the factory must cooperate, looking for potential problems and taking care of them in order to prevent future problems. Japan's strategy for dealing with these issues is the total quality control (TQC) system.

CONCLUSION

Every nation has the potential for achieving a more affluent society by introducing technology and developing added value in manufactured goods. India succeeded in supplying food for its population, and projections are that India will even export food in the near future. On the other hand, even today 60 percent of the world's population subsists at a starvation level. Therefore, there should be cooperation and understanding among nations concerning the use of technology as a tool for achieving an improved standard of living for all people. Unfortunately, the fruits of technology are often treated too politically to be used to upgrade the quality of human life. Nations should strive to introduce technology for the purpose of improving tomorrow's quality of life.

REFERENCES

Business Week. January 12, 1987. BW/Harris executive poll: Manufacturing's rise depends on the dollar, p. 68.

Meadows, D. H., D. L. Meadows, J. Randers, and W. W. Behrens III. 1972. The Limits to Growth. New York: Universe Books.

APPENDIXES

APPENDIX

A

The Council of Academies of Engineering and Technological Sciences

---

The Council of Academies of Engineering and Technological Sciences (CAETS),* an international nongovernmental association, was established in 1978 in recognition of the increasing dependence of nations for their economic and social welfare on technology and its advances. The Council's primary purpose is to promote international cooperation in engineering and technology in order to facilitate effective contributions of technological progress to the societies of all nations.

The objectives of CAETS are to provide an international forum for the discussion of technological and engineering issues; encourage cooperative international engineering efforts; foster the establishment of additional national engineering academies; and contribute to the strengthening of engineering and technological activities in order to promote economic and social welfare throughout the world.

A significant and distinguishing characteristic of CAETS is its unique criteria for membership. Member organizations are composed of members elected by peers on the basis of significant personal contributions to engineering, technological sciences, or related technological activities. Member organizations are governed by their elected memberships, and, by significant activity, demonstrate the compatibility of their objectives with those of the Council.

The Council's biennial convocations are the main mechanism for bringing together members of the international engineering community. Including the most recent convocation, six such meetings have been held.

---

*The organization was established in 1978 as the Convocation of Engineering Academies and Like Organizations. The current title was adopted in 1985.

- The First Convocation, hosted in 1978 by the National Academy of Engineering of the United States, was an opportunity for the participating academies to review their history, experiences, and guiding policies and to set objectives and program issues.
- The Second Convocation, hosted in 1980 by the Australian Academy of Technological Sciences, focused on the management of technological change.
- The Third Convocation, hosted in 1981 by the National Academy of Engineering of Mexico, examined engineering education in developed and developing nations.
- The Fourth Convocation, hosted in 1983 by the Royal Swedish Academy of Engineering Sciences, focused on technological trends in aviation, biotechnology, artificial intelligence, new materials, and automated production.
- The Fifth Convocation, hosted in 1985 by the Fellowship of Engineering of the United Kingdom, dealt with undersea engineering, food production, freight transport, telecommunications, and manufacturing in a global context.
- The Sixth Convocation, hosted in 1987 by the National Academy of Engineering of the United States, examined the impact of technology on the global economy.

APPENDIX

B

Sixth Convocation Council Member Academies and Officers

MEMBER ACADEMIES

Australian Academy of Technological Sciences and Engineering*
Clunies Ross House
Room 300
191 Royal Parade
Parkville, Victoria 3052
Australia

Danish Academy of Technical Sciences
266 Lundtoftevej
DK 2800 Lyngby
Denmark

The Fellowship of Engineering*
2 Little Smith Street
Westminster, London SW1P 3DL
United Kingdom

National Academy of Engineering of the United States of America*
2101 Constitution Avenue, N.W.
Washington, D.C. 20418
United States of America

National Academy of Engineering of Mexico*
Apartado Postal 20-733
Mexico 20, D.F.
Mexico

---

*Founding Members

Royal Swedish Academy of Engineering Sciences*
Box 5073
S-102 42 Stockholm 5
Sweden

OFFICERS

President:	Dr. H. Guyford Stever Foreign Secretary National Academy of Engineering of the United States of America
Vice President:	Sir David Zeidler President Australian Academy of Technological Sciences and Engineering
Secretary:	Steven N. Anastasion Consultant National Academy of Engineering of the United States of America

*Founding Members

APPENDIX

C

The Sixth Convocation Program: Technology and the Global Economy

---

Monday, March 30, 1987		**Auditorium**
4:00 pm	**Welcome and Opening Remarks**	H. Guyford Stever Council President
	Presidents of Council Member Academies	Sir David Zeidler President Australian Academy of Technological Sciences
		Poul Vermehren President Danish Academy of Technical Sciences
		Jaime Cervantes Vice President National Academy of Engineering of Mexico
		Hans G. Forsberg President Royal Swedish Academy of Engineering Sciences
		Sir Denis Rooke President The Fellowship of Engineering

		Robert M. White President National Academy of Engineering of the United States of America
	Introduction	Stephen D. Bechtel, Jr. Chairman Bechtel Group, Inc. United States
	Keynote	
	Globalization of Industry and Implications for the Future	Simon Ramo, Co-Founder TRW Inc., United States

Tuesday, March 31, 1987 **Lecture Room**

9:00 am	**Session 1: Cutting Edge Technologies and Their Effects** Chairman: Marco A. Murray-Lasso, National Academy of Engineering of Mexico	
9:15 am	Manufacturing Technologies	Pehr Gyllenhammar Chairman AB Volvo, Sweden
10:00 am	New Materials	Pierre Aigrain Scientific Advisor to the President of the Thomson Group, France
11:00 am	Information Technologies	Lars Ramqvist Executive Vice President Ericsson Group, Sweden
1:30 pm	**Session 2: Changes in Industry Sectors Resulting from Technology Advances** Chairman: Sir Francis Tombs, Chairman, Rolls-Royce, Ltd.; Director, Rothschilds, United Kingdom	
1:45 pm	Construction	Alden Yates President Bechtel Group, Inc. United States
2:30 pm	Software	George E. Pake Group Vice President Corporate Research (retired) Xerox Corp., United States

3:30 pm	Telecommunications	Hiroshi Inose Chairman Faculty of Engineering, University of Tokyo; Director General, National Center for Science Information Systems, Japan

Wednesday, April 1, 1987

9:00 am	**Session 3: Technology and the Global Economy: The Effects of Each on the Other** Panel Chairman: Alf Åkerman, Vice Chairman, Royal Swedish Academy of Engineering Sciences; Former President, SE-Banken, Gothenburg Group, Sweden	
9:15 am	Panel: Umberto Colombo, Chairman, Italian National Commission for Nuclear and Alternate Energy Sources, Italy Gerald Dinneen, Vice President, Research and Development, Honeywell Inc., United States Hajime Karatsu, Professor, Institute of Research and Development, Tokai University, Japan Ralph Landau, Consulting Professor of Economics and of Chemical Engineering, Stanford University, United States Enrique Martín del Campo, Executive Secretary for Education, Science and Culture, Organization of American States	
10:45 am	Discussion	
1:00 pm	**Session 4: Regional and National Consequences of Globalizing Industry** Chairman: Sir David Zeidler, President, Australian Academy of Technological Sciences, Australia	
1:10 pm	Latin America, Mexico, and the United States	Emilio Carillo Gamboa Director General, Telefonos de México, Mexico
1:50 pm	Western Europe	Harry L. Beckers Group Research Coordinator Shell International; Chairman, Industry Research & Development Advisory Committee, EEC, the Netherlands

2:45 pm	Pacific Rim	Jan E. Kolm Chairman National Energy Research Development and Demonstration Council (NRDDC), Former Director Imperial Chemical Industries (ICI) Australia Limited, Australia

4:00 pm **Session 5: Overall Assessment**
Panel Chairman: Morris Tanenbaum, Vice Chairman of the Board, AT&T, United States

4:15 pm Panel:
Wolf Häfele, Director General, Nuclear Research Center, Jülich, Federal Republic of Germany

Sir Robin Nicholson, Executive Director, Pilkington Brothers plc, United Kingdom

Robert Malpas, Managing Director, British Petroleum Company, United Kingdom

APPENDIX

D

List of Participants

AUSTRALIA

Malcolm Chaikin
Pro-Vice-Chancellor
The University of New South Wales

John L. Farrands
Chairman, International Relations Committee
Australian Academy of Technological Sciences

Jan E. Kolm
Chairman, National Energy Research Development and Demonstration Council
Former Director, Imperial Chemical Industries (ICI), Australia Limited

John C. Nixon
Australian Academy of Technological Sciences

Alan S. Reiher
Commissioner for North America
The Victorian Government of Australia

W. J. Stamm
President
The Institution of Engineers

W. J. McG. Tegart
Secretary, Department of Science
The Victorian Government of Australia

Robert G. Ward
General Manager
The Broken Hill Proprietary Co. Ltd.

Sir David Zeidler
President, Australian Academy of Technological Sciences
Vice President, CAETS

BELGIUM

René Bryssinck
President
National Institute of Research

A. Jaumotte
Vice President
Belgian Royal Academy

CANADA

Pierre Dupont
President
Bell Canada International, Inc.

COLOMBIA
Carlos Ospina
Senior Partner and Manager
Ingenieros Consultores Civiles y Electricos Ltda.

DENMARK
Poul Vermehren
President
Danish Academy of Technical Sciences

Vibeke Q. Zeuthen
Executive General Secretary
Danish Academy of Technical Sciences

FEDERAL REPUBLIC OF GERMANY
Dieter Behrens
Managing Director
Deutsche Gesellschaft für Chemisches Apparatewesen, Chemische Technik, und Biotechnologie eV

Wolf Häfele
Director General
Nuclear Research Center, Jülich

Klaus Pinkau
Director
Max Planck Institute of Plasma Physics

Franz Pischinger
Director, Institute for Applied Thermodynamics, Aachen Technical University
Vice President, German Research Society

FINLAND
Jorma Routti
Vice President and President-Elect
Finnish Academy of Technology

FRANCE
Pierre Aigrain
Scientific Advisor to the President of the Thomson Group

Andre Blanc-Lapierre
President
French Academy of Sciences

Alexis Dejou
President, Union Technique de l'Electricité
President, French Academy of Sciences

HUNGARY
Jenoe Fock
President, Federation of Technical and Scientific Societies
Former Prime Minister

Janos Ginsztler
Professor and Head, Department of Electrical Engineering Materials
Technical University of Budapest

Jozsef Hatvany
Advisor to the Director
Computer & Automation Institute
Hungarian Academy of Sciences

Mr. Lajos Nyiri
Science Counsellor
Hungarian Embassy
Washington, D.C.

ISRAEL
Shimon Yiftah
Professor
Department of Nuclear Engineering
Technion-Israel Institute of Technology

ITALY
Umberto Colombo
Chairman
Italian National Commission for Nuclear and Alternate Energy Sources

JAPAN

Hiroshi Inose
Chairman
The Faculty of Engineering, University of Tokyo
Director General, National Center for Science Information Systems

Hajime Karatsu
Professor
Institute of Research and Development
Tokai University

MEXICO

Eduardo Campero Littlewood
Secretary
National Academy of Engineering of Mexico

Emilio Carrillo Gamboa
Director General
Telefonos de México

Jaime Cervantes de Grotari
Vice President
National Academy of Engineering of Mexico

Abraham Charnes
Secretary, International Affairs
National Academy of Engineering of Mexico

Alejandro López-Toledo
Foreign Secretary
National Academy of Engineering of Mexico

Enrico N. Martínez
Universidad Autónoma Metropolitana-Iztapalapa

Alfredo Perez de Mendoza
Member
National Academy of Engineering of Mexico

Marco A. Murray-Lasso
Former President
National Academy of Engineering of Mexico

NETHERLANDS

H. L. Beckers
Director of Research, Shell International Research
President, Forum of Engineers

S. van Houten
Research and Development
Board of Management
Philips' Industries

NORWAY

Harald A. Øye
President
The Norwegian Academy of Technical Sciences

PEOPLE'S REPUBLIC OF CHINA

Yang Anxian
Deputy Chief, Planning Bureau Chinese Academy of Sciences
Visiting Scholar, Georgia Institute of Technology

Shi Changxu
Director, Technological Science Division, Chinese Academy of Sciences
Vice Chairman, National Natural Science Foundation of China

Teng Teng
Vice President, Chinese Academy of Sciences
President, CAS University of Science and Technology, Anhui

Zhang Guangdou
Professor, Vice President of University Council, Tsinghua University
Member of Presidium of Chinese Academy of Sciences

SPAIN

Jaime Sodupe
Director
Center for the Development of Industrial Technology

SWEDEN

Alf Åkerman
Vice Chairman, Royal Swedish Academy of Engineering Sciences;
Former President, SE-Banken—Gothenberg Group

Hans G. Forsberg
President
Royal Swedish Academy of Engineering Sciences

Pehr G. Gyllenhammar
Chairman
AB Volvo

Bengt Modeer
Executive Secretary
Royal Swedish Academy of Engineering Sciences

Lars Ramqvist
Executive Vice President
Ericsson Group

SWITZERLAND

Adrian W. Roth
Vice President
Swiss Academy of Engineering

UNITED KINGDOM

Geoffrey A. Atkinson
Deputy Secretary
The Fellowship of Engineering

Sir William Barlow
Senior Vice President, The Fellowship of Engineering
Chairman, BICC plc

Viscount Caldecote
Past President
The Fellowship of Engineering

Anthony R. Cox
Science Counselor
British Embassy, Washington, D.C.

John W. Fairclough
Chief Scientific Advisor
Cabinet Office

John R. Forrest
The Fellowship of Engineering

Robert Malpas
Managing Director
The British Petroleum Company

J. R. S. Morris
Vice President
The Fellowship of Engineering

Sir Robin Nicholson
Executive Director
Pilkington Brothers plc

B. W. Oakley, CBE
Director
The Alvey Directorate

V. J. Osola
Executive Secretary
The Fellowship of Engineering

G. K. C. Pardoe
Chairman
The Watt Committee on Energy Ltd.

Geoffrey Pattie
Minister of Industry and Information Technology
Department of Trade and Industry

Sir Denis Rooke
President, The Fellowship of Engineering;
Chairman, British Gas, plc

Sir Francis Tombs
Chairman, Rolls-Royce Ltd.
Director, Rothschilds

Peter O. Wolf
The Fellowship of Engineering

UNITED STATES

Steven N. Anastasion
Consultant, National Academy of Engineering
Secretary, CAETS

Holt Ashley
Professor, Department of Aeronautics and Astronautics
Stanford University

Isaac L. Auerbach
President
Auerbach Consultants

Jesse H. Ausubel
Director, Program Office
National Academy of Engineering

Jordan J. Baruch
President
Jordan Baruch Associates

Stephen D. Bechtel, Jr.
Chairman
Bechtel Group, Inc.

Daniel Berg
Institute Professor of Science and Technology
Rensselaer Polytechnic Institute

Erich Bloch
Director
National Science Foundation

W. Dale Compton
Senior Fellow
National Academy of Engineering

Dale R. Corson
President Emeritus
Cornell University

Gerald P. Dinneen
Vice President
Science and Technology
Honeywell Inc.

John V. Evans
Vice President for R&D and Director
COMSAT Laboratories
COMSAT Communications Satellite

Daniel J. Fink
President
D. J. Fink Associates, Inc.

Alexander H. Flax
Home Secretary
National Academy of Engineering

Edward R. Kane
Director and Former President
E. I. du Pont de Nemours & Company

Harold Liebowitz
Dean
School of Engineering and Applied Science
The George Washington University

The Honorable D. Bruce Merrifield
Assistant Secretary for Productivity, Technology and Innovation
U.S. Department of Commerce

Hugh Miller
Consultant
National Academy of Engineering

John H. Moore
Deputy Director
National Science Foundation

Lt. General John W. Morris
President and CEO
Engineering Group
Planning Research Corporation

Peter Murray
Director
Nuclear Programs
Westinghouse Electric Corporation

George E. Pake
Group Vice President
Corporate Research (retired)
Xerox Corporation

Frank Press
President
National Academy of Sciences

Walter A. Rosenblith
Emeritus Institute Professor
Massachusetts Institute of Technology

William C. Salmon
Executive Officer
National Academy of Engineering

Philip M. Smith
Executive Officer
National Academy of Sciences

H. Guyford Stever
Foreign Secretary, National Academy of Engineering
President, CAETS

Morris Tanenbaum
Vice Chairman of the Board
AT&T Corporate Headquarters

Gerald F. Tape
Retired President
Associated Universities, Inc.

Alvin W. Trivelpiece
Director
Office of Energy Research
U.S. Department of Energy

Albert R. C. Westwood
Corporate Director
Research & Development
Martin Marietta Corporation

Robert M. White
President
National Academy of Engineering

F. Karl Willenbrock
Executive Director
American Society for Engineering Education

Alden P. Yates
President
Bechtel Group, Inc.

USSR

Konstantin V. Frolov
Vice President
Academy of Sciences of the USSR

V. Smirnyagin
Director
Foreign Relations Section
Academy of Sciences of the USSR

ORGANIZATION OF AMERICAN STATES (OAS)

Enrique Martín del Campo
Washington, D.C.

UNION OF INTERNATIONAL TECHNICAL ASSOCIATIONS (UITA)

Leo S. Packer
France

WORLD FEDERATION OF ENGINEERING ORGANIZATIONS (WFEO)

Michael W. Leonard
France

APPENDIX

E

Contributors

PIERRE AIGRAIN is scientific advisor to the president of the Thomson Group. Trained in physics at the Faculty of Sciences of the University of Paris and at the Carnegie Institute of Technology in the United States, he spent the early portion of his career as a professor of the Faculty of Sciences in Paris. Dr. Aigrain's work there on electrotechnology and energy led to a lasting association with the French Atomic Energy Commission. He has held key positions with the French government, including that of minister of state for research from 1978 to 1981. Dr. Aigrain is the author of numerous scientific papers on electronic circuits and semiconductors, has served as both the secretary general and vice president of the French Physical Society, and was chairman of the French-Chinese Committee for Economic and Industrial Cooperation.

HARRY L. BECKERS is chairman of the Industry Research and Development Advisory Committee of the European Economic Community and group research coordinator for Shell International. Dr. Beckers was also president of the European Industrial Research Association, president of the Forum of Engineers of the Netherlands, and a member of the Dutch Advisory Committee on Science Policy to the Dutch government.

EMILIO CARRILLO GAMBOA is Ambassador of Mexico to Canada. He formerly served as director general and chief executive officer for Telefonos de México. In his capacity as director general, Mr. Carrillo Gamboa was chairman of the board of directors of Telefonos de México's 23 affiliates. In addition, he is a member of the board of several private and public enterprises in Mexico. Mr. Carrillo Gamboa maintains an active schedule as a lecturer for universities, civic associations, and public and private institutions. He is a graduate of law of the Universidad Nacional Autónoma de México and did graduate work at Georgetown University in Washington, D.C.

UMBERTO COLOMBO is chairman of the Italian National Commission for Nuclear and Alternative Energy Sources. He has also served as director of the G. Donegani Research Center, director general for research and corporate strategies of Montedison Company, and chairman of the Italian Atomic Energy Commission. Dr. Colombo is chairman of the European Economic Community's Committee on Science and Technology. Dr. Colombo received his doctorate in physical chemistry from Pavia University and studied under a postdoctorate Fulbright Fellowship at the Massachusetts Institute of Technology.

GERALD P. DINNEEN is vice president of science and technology for Honeywell Inc. In this capacity, he manages the corporate technical centers and assists corporate, group, and division management in areas of technology and engineering. After receiving his Ph.D. in mathematics from the University of Wisconsin, Dr. Dinneen joined the Lincoln Laboratory at the Massachusetts Institute of Technology where he held a series of positions including the directorship from 1970 to 1977. From 1971 to 1981, he was also professor of electrical engineering at MIT. In 1977, Dr. Dinneen was named Assistant Secretary of Defense for Communications, Command and Control, and Intelligence, a position he held until he joined Honeywell in 1981. Dr. Dinneen is a member of the board of directors of the Microelectronics and Computer Technology Corporation (MCC), the Corporation for Open Systems, and Honeywell-NEC Supercomputers.

PEHR GYLLENHAMMER is chairman and chief executive officer of AB Volvo. He began his career with the Ampion Insurance Company and later joined the Skandia Insurance Company where he became deputy managing director and managing director and chief executive. He joined Volvo in 1971 as managing director and chief executive officer, assuming his current position in 1983. In addition to his position with Volvo, Mr. Gyllenhammar is a member of numerous corporate boards, a member of the board of the Swedish Employers' Confederation, the Federation of Swedish Industries, and a member of the International Advisory Committee of the Chase Manhattan Bank. Mr. Gyllenhammar received his bachelor of law degree from Lund University and pursued further study in Great Britain, the United States, and Switzerland.

WOLF HÄFELE is director general of the Nuclear Research Center, Jülich/FRG. He has held numerous positions in the field of nuclear energy physics at universities and research institutes, including head of the nuclear safeguards project and scientific adviser to the government of the Federal Republic of Germany on the nonproliferation treaty, head of the energy research systems project at the International Institute for Applied Systems Analysis (IIASA), and deputy director of IIASA. He has been awarded several honors for his work in this area, including the Federal Service Cross and the Austrian Cross of Honor for Science and Arts. Dr. Häfele received his Ph.D. in theoretical physics from the University of Göttingen.

HIROSHI INOSE is director general of the National Center for Science Information Systems (Japan) and the dean of the Faculty of Engineering at the University of Tokyo. Since 1984, Dr. Inose has also been chairman of the Committee for Scientific and Technological Policy of the Organization for Economic Cooperation and Development. He is a fellow of the Institute of Electrical and Electronics Engineers and a foreign associate of both the National Academy of Engineering and the National Academy of Sciences. Dr. Inose is widely known for his work in digital communication and road traffic control. Dr. Inose received his bachelor's and doctorate degrees in engineering from the University of Tokyo.

HAJIME KARATSU is a professor at the R&D Institute of Tokai University. Following his graduation with a degree in electrical engineering from the University of Tokyo, Professor Karatsu began working for Nippon Telegraph and Telephone Public Corporation. He later joined Matsushita Communication Industrial Company, Ltd., where he assumed the posts of director, managing director, and technical adviser to Matsushita Electric Industrial Company. Professor Karatsu has served on the Industrial Structural Council of the Ministry of International Trade and Industry, with the Institute of Fifth Generation Computer Systems, and as chairman of the Office Automation Committee of the Tokyo metropolitan government. For his contributions in the field of statistical quality control, Professor Karatsu received Japan's Deming Prize in 1981.

JAN E. KOLM is chairman of the National Energy Research Development and Demonstration Council and former executive director of Imperial Chemical Industries (ICI) Australia Ltd. During his career with ICI Australia, Mr. Kolm held the posts of corporate research manager, technical and research director, and director of ICI Engineering, CSR Chemicals, and Nylex Corporation. He was involved in the development of electrolytic cells, nylon intermediates, a process for hexachlorocyclohexane, and a new route to tetraisole, a veterinary drug licensed worldwide. Mr. Kolm has also served as a member of the advisory committee to the Commonwealth Scientific and Industrial Research Organization (CSIRO) and as chairman of the Victorian

State Committee of CSIRO. Mr. Kolm holds a degree in chemical engineering from Prague Technical University.

RALPH LANDAU is former chairman of the board of the Halcon SD Group, Inc., a high-technology business in the chemical industry. When his interests were sold in 1982, he became consulting professor of economics at Stanford University, and subsequently a fellow of the faculty of the Kennedy School at Harvard. He is a trustee of the Massachusetts Institute of Technology, the California Institute of Technology, the University of Pennsylvania, and the Cold Spring Harbor Laboratory, as well as a retired director of ALCOA. He is vice president of the National Academy of Engineering and in 1985 was among the first recipients of the National Medal of Technology. He holds an Sc.D. from MIT.

ROBERT MALPAS is managing director of the British Petroleum Company. He began his career at Imperial Chemical Industries where he held numerous positions during his 30-year tenure there. In 1978 he became president of Halcon International Inc. He assumed his current position in 1983. Mr. Malpas is a fellow of the Institute of Chemical Engineers, the Institute of Materials Handling, and the Institution of Mechanical Engineers. He is affiliated with the British Oxygen Group, the Advisory Council for Applied Research and Development (ACARD), and the Engineering Council. Mr. Malpas received his degree from Durham University.

ENRIQUE MARTÍN DEL CAMPO is assistant secretary for education, science, and culture of the Organization of American States. He has extensive experience in international technical cooperation, particularly in the Americas. He has also served as minister counselor for scientific and technological affairs at the Mexican Embassy in Washington, D.C., consultant to the United Nations, research engineer at L. M. Ericcson of Sweden, and director of international affairs for the National Council for Science and Technology of Mexico. Dr. Martín del Campo is a founding member of the National Academy of Engineering of Mexico. He holds a doctorate in physics and electronics from the University of Toulouse, France.

JANET H. MUROYAMA is program associate at the National Academy of Engineering of the United States. She works on the Academy's international programs and coordinates a project on the human resource and organizational aspects of the adoption of new workplace technologies.

SIR ROBIN NICHOLSON is an executive director of Pilkington Brothers plc. Sir Robin has taught in the departments of metallurgy at both Cambridge and Manchester universities. In 1972 he joined Inco Europe Ltd. as director of the company's research laboratory and later assumed the positions of director and managing director of the company. From 1981 to 1985, Sir Robin served in the Cabinet Office, first as chief scientist of the central policy review staff and then as chief scientific adviser. He is a non-executive director of Rolls Royce and a fellow of the Royal Society and the Fellowship of Engineering. He has a Ph.D. in metallurgy from Cambridge University.

GEORGE E. PAKE recently retired from Xerox Corporation where he was group vice president for corporate research. After receiving his Ph.D. degree in physics from Harvard University in 1948, he joined the faculty of Washington University in St. Louis. In 1956 he became professor of physics at Stanford University. Dr. Pake returned to Washington University as provost and executive vice chancellor in 1962. In 1970 he joined Xerox Corporation to establish the Xerox Palo Alto Research Center. He is now director of the Institute for Research on Learning, a new nonprofit research institute established under a grant from Xerox Corporation.

SIMON RAMO, recipient of the Presidential Medal of Freedom and the National Medal of Science, is the "R" of TRW Inc. The chief scientist in developing the U.S. intercontinental ballistic missile, he was chairman of the President's Committee on Science and Technology under President Ford, a member of the Advisory Council to Secretary of State Kissinger on Science and Foreign Affairs, the White House Council on Energy Research and Development, and the National Science Board and the first to receive the National Academy of Engineering's award for statesmanship in national science and technology policy. Dr. Ramo also is the author of widely used textbooks in science, engineering, and management.

LARS RAMQVIST is executive vice president of the Ericsson Group. He began his career as head of the materials laboratory at Stora Kopparberg and later conducted research at Axel Johnsson Institute, of which he became president. In 1980, he joined Ericsson, assuming the positions of vice president of information systems, head of strategic planning, and senior vice president. In his current position he has overall group responsibility for technology, product and production techniques, and quality assurance systems. Dr. Ramqvist studied at the Universities of Uppsala and Stockholm and holds a Ph.D. in solid-state physics and chemistry.

NATHAN ROSENBERG is Fairleigh S. Dickinson, Jr. Professor of Public Policy at Stanford University. Before moving to Stanford in 1974, he served on the faculty of the University of Wisconsin, Harvard University, Purdue University, and the University of Pennsylvania. He has written extensively on the economics of technological change. His most recent books are *The*

Positive Sum Strategy, *Inside the Black Box*, and *How the West Grew Rich*. Dr. Rosenberg earned his M.A. and Ph.D. degrees from the University of Wisconsin.

H. GUYFORD STEVER, foreign secretary of the National Academy of Engineering of the United States, has spent his career as a scientist, engineer, educator, and administrator. He was professor of aeronautics and astronautics at the Massachusetts Institute of Technology for 20 years as well as head of the departments of mechanical engineering, and naval architecture and marine engineering. From 1965 to 1972, Dr. Stever was president of Carnegie Mellon University, during which time the Carnegie Institute of Technology and the Mellon Institute were merged. Dr. Stever was director of the National Science Foundation as well as Science Advisor to the President from 1972 to 1976. Subsequently, he served as White House Science and Technology Advisor to President Ford and director of the Office of Science and Technology Policy. Dr. Stever has a Ph.D. in physics from the California Institute of Technology.

MORRIS TANENBAUM is vice chairman of the board of AT&T, responsible for finance and planning. As a member of the technical staff of Bell Telephone Laboratories, he invented the process for the first practical silicon transistors. Dr. Tanenbaum has held several executive positions throughout AT&T, including vice president of engineering at Western Electric Company, executive vice president of Bell Laboratories, president of New Jersey Bell Telephone Company, and first chairman and chief executive officer of AT&T Communications. He is a trustee of Johns Hopkins University, Massachusetts Institute of Technology, Battelle Memorial Institute, and the Brookings Institution and serves on the board of directors of a number of companies. Dr. Tanenbaum received his Ph.D. in physical chemistry from Princeton University.

ALDEN P. YATES is president and chief operating officer of Bechtel Group, Inc. Mr. Yates has spent his career designing, constructing, and managing complex engineering projects. In his current position, he is responsible for the day-to-day management of Bechtel projects worldwide. These projects, serving many industries and governments, face a variety of technical challenges ranging from deep sea structures to outer space. Mr. Yates is a civil engineering graduate of Stanford University and serves on the advisory council to the School of Engineering.

Index

D

E

F

G

H

N

O

P

Q

R

S

W

X